Genetic Research in Psychiatry

"2. Münchner Genetikgespräche"
September 12 - 15, 1991
C.I.N.P. President's Workshop

Editors
J. Mendlewicz H. Hippius

Co-Editors
B. Bondy M. Ackenheil M. Sandler

With 17 Figures and 25 Tables

Springer-Verlag
Berlin Heidelberg New York
London Paris Tokyo
Hong Kong Barcelona
Budapest

Editors:

Professor Dr. Julien Mendlewicz
University Hospital of Brussels
Hôpital Erasme
Route de Lennik 808
B-1070 Bruxelles
Belgium

Professor Dr. Hanns Hippius
Head of Dept. of Psychiatry
University of Munich
Nussbaumstrasse 7
D-W-8000 München 2
Germany

Co-Editors:

PD Dr. Brigitta Bondy
Dept. of Psychiatry
University of Munich
Nussbaumstrasse 7
D-W-8000 München 2
Germany

Professor Dr. Manfred Ackenheil
Dept. of Psychiatry
University of Munich
Nussbaumstrasse 7
D-W-8000 München 2
Germany

Professor Dr. Merton Sandler
Department of Chemical Pathology
Queen Charlotte's Maternity Hospital
Goldhawk Road, London W6 OXG
United Kingdom

ISBN-13: 978-3-642-46764-6 e-ISBN-13: 978-3-642-46762-2
DOI: 10.1007/978-3-642-46762-2

Production: Congress Project Management GmbH, Letzter Hasenpfad 61, D-W-6000 Frankfurt am Main 70,

25/3130-5 4 3 2 1 0 – Printed on acid-free paper

Preface

The field of psychiatric genetics was given scientific impetus by developments in molecular biology in recent years. Almost coinciding with the appearance of the first, very exciting results on linkage analyses in affective disorders, an international symposium was organized in collaboration between German and American investigators as the first of the *Münchener Genetik-Gespräche* and held in Berlin in autumn 1986. Now, 5 years after the first workshop and after a series of rather promising but also disappointing results, the second of these meetings was organized in collaboration with the C.I.N.P., the C.I.N.P. President's Workshop Genetic Research in Psychiatry held in Prien, Bavaria, FRG, from 12-15 September 1991. The aim of this conference was to provide a forum for the review and discussion of recent developments not only in molecular biology but also for the characterization of psychiatric phenotypes. Thus basic and clinical research issues were addressed and the focus was inherently interdisciplinary, providing a review and high-quality update of scientific knowledge by experts from all over the world in the field of psychiatric genetics.

The exponential progress in the field of molecular biology, the almost daily increasing number of cloned and sequenced genes which are potentially important in psychiatric disturbances, urges close international collaboration and mutual exchange of data and methods. Important initiatives are growing up, not only in the United States. The network of the European Science Foundation and in Germany the research project "Genetics in Psychiatry" sponsored by the Deutsche Forschungsgemeinschaft are good examples of increasing joint initiatives to characterize susceptibility genes for mental disorders.

This workshop bridged the gap between molecular biology and clinical research, as it underlined the complex and heterogeneic nature of behavioral neuropsychiatric disorders and the necessity to develop new mathematical methods of analyses. Above all, innovative instruments for phenotypical characterization are needed to further advance our knowledge of the genetics of complex behavioral diseases.

We wish to express our gratitude and appreciation to the speakers and to all the other persons who contributed their time and effort to make the conference a success and challenge for the future.

January 1992

Julien Mendlewicz
Hanns Hippius

Contents Page

Molecular Genetics of Affective Psychoses*

M. Baron

Introduction

Studies with chromosomal markers constitute a powerful strategy for demonstrating major gene effects. Recently reported linkages between affective disorders and specific genomic regions have attracted considerable attention in the psychiatric and genetic literature (Baron et al. 1987; Egeland et al 1987; Mendlewicz et al. 1987). However, a string of conflicting data has cast doubt on some of these findings and has underscored the problems inherent in studying these complex diseases (reviews: Baron et al. 1990a; Risch 1990a).

In this article the methodological and conceptual issues as they bear upon the recent linkage results are discussed. Then guidelines are offered for enhancing the prospects of genetic mapping efforts in this area of enquiry. I hope to provide a viewpoint that will be useful both for researchers involved in this particular discipline as well as for scientists and scholars outside this field who might be interested in human disorders with complex inheritance. Some of these issues have been considered elsewhere (Edwards and Watts 1989; Baron 1990a, 1991a; Baron et al. 1990a,b; Elston and Wilson 1990; Green 1990; Plomin 1990; Risch 1990a; Saurez et al. 1990; Baron 1990b).

Background

Affective disorders are common behavioral conditions of unknown etiology which affect mood, cognition, and perception. They can lead to impaired psychosocial functioning, multiple hospitalizations, drug and alcohol abuse, and suicide. With a combined lifetime prevalence of 10%-15% in all ethnic groups, they are among the most common disorders in humans. Although the efficacy of somatic therapies for these disorders is well established, it has not been possible to define consistent neurobiological abnormalities of etiologic significance.

Family, twin, and adoption studies support the role of heredity in affective psychoses; however, little is known about the underlying genetic mechanisms. Segregation analysis and related statistical methods have not resolved the mode of

*This work was supported by NIMH Research Scientist Development Award MH00176 and by grants MH42535, MH44115, and MH43979.

transmission; the inheritance pattern is complex and does not follow mendelian expectations (Baron 1991b). Early studies with classical chromosomal markers such as leukocyte and erythrocyte antigens and serum proteins yielded negative or otherwise inconclusive results. For the most part, those studies were undermined by methodological drawbacks concerning diagnosis, ascertainment of pedigrees, and statistical analysis, as well as dearth of polymorphic markers (Gershon 1990; Baron 1991b).

Recent advances in diagnostic procedures, molecular genetic techniques (particularly the identification of polymorphic DNA markers that span the human genome), and statistical methods for linkage analysis have revived the search for disease-related genes, leading to a new harvest of data. Since 1987, several reports of linkage between gene markers and affective disorders have appeared. Bipolar and related affective disorders were reported linked to two chromosome 11p marker loci, the Harvey-RAS (HRAS1) and insulin, in an Old Order Amish pedigree (Egeland et al. 1987). Linkage was also reported to the Xq28 markers color blindness and glucose-6-phosphate dehydrogenase (G6PD) in Israeli families (Baron et al. 1987) and to the Xq27 factor IX (F9) locus in a Belgian population (Mendlewicz et al. 1987).

Unfortunately, other research groups have failed to reproduce the chromosome 11p and chromosome Xq27-28 linkages (reviews: Baron et al. 1990a; Risch 1990a; Baron 1991b,c). Moreover, when the Amish data (Egeland et al. 1987) were extended and reanalyzed, the evidence of linkage to chromosome 11p was substantially diminished, casting doubt on the original finding (Kelsoe et al. 1989). The negative linkage data have tempered the euphoria generated by the early findings. Although some of the discordant results can be attributed to etiologic heterogeneity, there is a distinct possibility of spurious linkage findings owing to methodological faults and other uncertainties in studying complex disorders. These issues and their implications are discussed in the following sections.

Methodological Issues

The Phenotype

Despite advances in diagnostic practices and improved reliability and consensus among clinicians and investigators (Spitzer et al 1978; DSM-III-R 1987) the definition of the affective phenotype remains a thorny issue. In the absence of external validating measures (e.g., biological correlates), the diagnosis rests primarily on behavioral signs and symptoms, thus placing limits on its validity. Other concerns are noteworthy. Disorders thought to comprise a particular disease phenotype, based on familial aggregation, tend to range from rare, severe clinical states to relatively common, milder conditions. The latter are likely more heterogeneous and may include misclassified cases also known as phenocopies or false positives, that is, those erroneously considered affected. Specifically, the disease phenotype extends from manic depression or bipolar I disorder (mania in association with depression), which represents the severe end of the spectrum, to milder clinical variants, such as bipolar II disorder (hypomania and depression) and unipolar depression (depression only). The question of the phenotype may be further complicated by clinical states that cross the boundaries between the major psychoses (e.g., schizoaffective illness, which combines both schizophrenic and affective symptoms, and some depressive states). Another issue to contend with is phenotypic variation within diagnostic categories. For instance, the range of symptoms, age at onset, and illness course vary greatly among patients in the various diagnostic categories. The diagnostic criteria in the conventional classification schemes do not adequately address this variability. Also, the quantity and quality of the clinical information may vary among individuals, i.e., some diagnoses are based on systematic and internally consistent observations encompassing personal interviews, information obtained from other relatives, and medical records, whereas other diagnoses may be compromised by contradictory or otherwise limited clinical information. Finally, disparate diagnostic outcomes among research teams using the same diagnostic procedures have been reported (Zimmerman 1988), and diagnostic "drift" within the same research group is known to occur. This is likely a reflection of the uncertainties inherent in the clinical diagnosis. These uncertainties can lead to another type of phenotypic misclassifications called false negatives, that is, those misclassified as unaffected. To be sure, in some cases the appearance of a "normal" phenotype can be ascribed to incomplete penetrance (incomplete manifestation of the disorder in subjects who have the abnormal genotype).

Mode of Inheritance

The common statistical method for linkage analysis assumes single locus inheritance. However, this assumption may not be tenable in the absence of prior evidence to that effect (e.g., segregation analysis, biochemical clues). That is not to say that single major genes may not operate in some subsets of mental disorders. But the genetic parameters (i.e., penetrance, allele frequency) which must be specified in the calculation of the lod score (the logarithm of odds in favor of the linkage hypothesis) will not be known with certainty. In some situations, misspecification of genetic parameters may reduce the expected lod score (Clerget-Darpoux et al. 1986). However, the use of multiple models may result in the opposite outcome, namely, inflation of the lod score (Weeks et al. 1990). This is a realistic scenario since different sets of parameters are often used in an attempt to capture the "true" genetic configuration.

It may also be that more than one locus is involved in etiology, i.e., interaction of alleles at different loci (epistasis). In this circumstance, it may still be possible to detect the contributing loci by linkage analysis assuming a single mendelian locus (Majumder 1989). This, however, may result in a deflated lod score. This is an important consideration because epistasis may be the true state of nature for some mental disorders. For instance, it has been observed that published prevalence rates for schizophrenia best fit an epistatic model with two or three major loci (Risch

1990b). The persistence of this illness in the population despite significant selective disadvantage (i.e., reduced reproduction) also argues against the general presence of single locus inheritance. Restricting the analysis to affected sibs (sib-pair method) or other affected relatives circumvents some of these problems because, unlike the lod score approach, no assumptions are required about mode of inheritance parameters; however, this method is considered less efficient and less powerful than the lod score approach (Bishop and Williamson 1990). Methods are not yet available for the detection of common alleles, each of small effect, in the event of multiple minor locus contributions to etiology.

Etiologic Heterogeneity and Sample Design

Affective disorders are considered etiologically heterogeneous. Simulation studies have shown correspondence between the degree of heterogeneity and the sample size required to detect linkage (Ott 1986; Cavalli-Sforza and King 1986; Martinez and Goldin 1989a, b; Goldin 1990; Leboyer et al. 1990; Risch 1990c). For instance, if the proportion of linked cases is small (say, less than 20%), the required sample must be exceedingly large lest linkage be missed entirely. One solution to this problem is to study large pedigrees, each sufficiently informative to display linkage independently of other family data. Computer simulation of the expected lod score, given the phenotype at the disease locus, could be used to select pedigrees according to their suitability for linkage analysis (Ott 1989; Sandkuyl and Ott 1989). However, larger pedigrees may incur more opportunities for affected persons marrying in who may introduce a new genetic type of the illness into the pedigree (assortative mating). Also, phenocopies (especially among the common phenotypes) are more likely to occur by virtue of the sheer number of those affected. These circumstances would weaken the potential to detect linkage.

Sib-pair analysis may complement the pedigree analysis approach. However, this method, too, may be greatly compromised by the occurrence of phenocopies (Bishop and Williamson 1990). Alternatively, homogeneous population isolates with large dense pedigrees can be sought. Such isolates are rare, and the generality of the linkage results may be dubious.

Some investigators advocate systematic ascertainment of large populations (as opposed to targeting potentially promising pedigrees), followed by segregation analysis, as a prelude to linkage analysis. Presumably this could lead to insights into the genetic underpinnings of the disorder in question, specifically, the estimation of genetic parameters. Other experts, however, doubt the merit of this approach owing to the complex inheritance pattern, previous failures of segregation analysis to elucidate mode of inheritance, and doubts as to whether the presumed added precision in genetic parameter estimates is worth the cost involved. (Ott 1986; Baron et al. 1990a). At any rate, should linkage be found, the parameter set might be varied within certain limits to study its robustness. It is also possible to combine segregation and linkage analysis once linkage is detected.

The pros and cons of the various sampling strategies need to be considered in the design and interpretation of these studies. Statistical methods that exploit the power of a complete genomic linkage map could be used to reduce the sample needed to detect heterogeneous linkage (Lander and Botstein 1986), as do clinical and/or biological correlates that sort out potentially homogeneous disease subsets and reduce diagnostic uncertainties. A case in point is coronary heart disease, in which defects in cholesterol metabolism have led to the identification of specific gene effects. Another example is diabetes mellitus, in which clinical (early onset) and biological (insulin dependence) factors jointly identify a form of the illness that is associated with certain HLA haplotypes.

Genes and Environment

That the environment can impinge on the pathway from genotype to phenotype is widely accepted. Although specific environmental contributions to the etiology of affective psychoses are yet to be discovered, gene–environment interaction can be surmised from the incomplete monozygotic twin concordance for these disorders (akin to incomplete penetrance) and cohort effect (the increased rate of the disorder over time, or across generations) (Baron 1991b). Sporadic cases or phenocopies (i.e., environmentally caused disease unrelated to the particular genotype being studied) can also occur. Generally, these phenomena could mask the presence of linkage by giving rise to misspecified genetic parameters and phenotypic uncertainties. Appropriate statistical adjustments need to be considered in the linkage analysis (Baron et al. 1990a). Methods of incorporating environmental risk factors in linkage analysis are being developed.

Gene Markers

The uncertainties in conducting linkage studies without a prior "clue" to the underlying biological substrate have led some investigators to favor "candidate" loci (i.e., genes coding for products with neurobiological function, such as neuroreceptor proteins and enzymes; psychoactive drug receptors; brain mRNA that distinguishes diseased from normal tissue) over a systematic search throughout the genome using all available polymorphic markers. Clues to gene locations can also be obtained from cytological abnormalities such as deletions or translocations which assort with the illness. However, studies involving pivotal candidate genes are yet to yield positive results (Gurling 1985; Baron and Rainer 1988; Gershon 1990; Baron 1991b). The large proportion of human genes expressed in brain (approximately 50%), together with extraneous factors affecting postmortem tissues, present further difficulties with this approach. Likewise the search for biological correlates of genetic susceptibility (endophenotypes) has not led to consistent findings (Gershon 1990; Baron 1991b). Also, some potential endophenotypes may be too removed from the disease genotype to fulfill their promise. Finally, mental illness associated with chromosomal abnormalities could merely represent a nonspecific effect of the chromosomal aberration rather than a bona fide genetic disorder. For these reasons, an all-out mapping effort using the ever-improving genomic map might be more fruitful (albeit more expensive) than a narrow scope endeavor centered on candidate genes. The highly polymorphic minisatellite (variable number tandem repeats, or VNTRs) and microsatellite (CA repeat) DNA markers, which represent an improvement over the "first generation" DNA markers, will likely expedite this effort. Also, given the likely heterogeneity of mental disorders and the possible interaction among loci, the availability of extensive mapping information should give rise to a better understanding of the various factors that contribute to etiology. That is not to say that potential clues to putative disease genes should not be pursued concomitantly. A case in point are two potential candidate genes, the genes encoding the D_2 dopamine receptor (DRD2) and tyrosinase, both mapped on chromosome 11q, a genomic region with reported translocations segregating with bipolar illness and other psychoses (Smith et al. 1989; St. Clair et al. 1990; Holland and Gosden 1990). Another intriguing finding is the localization to chromosome Xq28, a region of interest in X-linkage studies of bipolar illness, of the gene encoding a subunit of the neurotransmitter γ-aminobutyric acid (GABRA3) (Buckle et al. 1989). Also of interest is the mapping of the hydroxyindole-o-methyltransferase (HIOMT) gene to the pseudoautosomal region of the sex chromosome, a genomic region implicated in bipolar illness and other psychoses (Crow 1988). HIOMT is a potential candidate gene in view of reported melatonin abnormalities in affective illness. Although neither of these genes appears linked to the illness in some pedigrees (Holmes et al. 1991; van Broeckhoven and Mendlewicz 1991; Robb et al. 1991), further study would be of interest.

The Lod Score Criterion

A lod score of 3 (corresponding to an odds ratio of 1000 to 1), conventionally used as the criterion for acceptance of linkage, may not be appropriate for mental disorders. The lod score statistic was devised specifically for single gene disorders. The absence of prior clear evidence for single locus inheritance, coupled with the other methodological issues discussed herein, requires caution in adopting the conventional lod threshold as "proof" of linkage. For example, the aforementioned uncertainties concerning the phenotype and the mode of inheritance often lead to maximization of the lod score over several diagnostic schemes and model parameters. Similarly, a false impression of linkage may occur due to fortuitous segregation when multiple genetic markers are examined. Given these complications, the critical lod score is unknown, and the traditional lod threshold may be marginal at best (Edwards and Watts 1989; Weeks et al. 1990). Methods for calculating the probability of falsely concluding linkage on the basis of a statistical test could complement and bolster the lod score statistic.

From Marker to Gene

Current molecular genetic techniques require a large number of informative matings (meiotic events) to generate a high resolution map of the region suspected to contain the putative disease gene. Namely, to characterize the region of interest, it is necessary to identify crossovers between the disease locus and DNA markers not more than several centimorgans away. Barring a fortuitous chromosomal aberration that could expedite this effort, such maps are essential for the isolation of the gene. Rare linked forms of mental disorders would make it difficult to obtain the amount of pedigree material required for the cloning procedures. Mapping efforts can also be hampered by ambiguous phenotype definitions, leading to large confidence intervals around the disease locus. New methods, for example, new cloning strategies, rich gene data banks, and improved disease classification, will have to be devised to address these problems.

True Versus False Linkage

The methodological issues discussed in the preceding section can lead to two categories of error: false positive, or type I error (a false claim that linkage exists), and false negative, or type II error (rejection of linkage when linkage is actually present).

Type I Errors

Type I errors can occur under the following circumstances:

1. Biased assignment of genotypes and/or phenotypes due to non-blind procedures for clinical diagnosis and genetic marker analysis.

2. Multiple test effects arising from testing a range of genetic models and diagnostic schemes, as well as various independent marker systems. This is a likely scenario in disorders with unknown genetic parameters and unclear phenotypic boundaries. Recently, the impact of multiple tests on linkage results has been assessed by simulation studies. Weeks et al. (1990) have examined the effect of maximizing the lod score over genetic model parameters (e.g., penetrance, allele frequency) and disease definitions. They have shown that in some instances the observed lod score can be inflated by as much as 1.5 units. Similarly, Clerget-Darpoux et al. (1990) and Goldin (1990), using a complementary approach, have found that the probability of a chance finding with multiple markers under various genetic models and diagnostic schemes is not negligible. For example, the probability of exceeding a lod score of 3 under independent segregation of marker and disease is almost 10% for linkage analysis performed with four penetrances, four diagnostic classifications, and 80 "marker-laboratory" tests (markers typed independently in different laboratories); this probability increases rapidly with the number of markers tested, a likely scenario in extensive genomic scans. (Clerget-Darpoux et

al. 1990). These results echo earlier suggestions concerning the need to deflate lod scores in multiple test situations (Thompson 1984; Ott 1986; Edwards and Watts 1989).

3. Selective presentation of families with positive lod scores to the exclusion of those with no evidence of linkage. The summation of these lod scores could then produce a spurious linkage finding. This problem can be especially acute with small positive scores because large individual lod values, which are less susceptible to ascertainment bias, could signal genuine linkage heterogeneity.

4. False inference of linkage based on borderline lod scores. In view of the methodological problems discussed herein, the conventional lod threshold may be merely suggestive. This is not to say that modest positive lod scores need not be pursued as potential "leads" for putative disease genes.

Errors occurring under the first three scenarios may be hard to detect. For example, although the importance of blindness in linkage studies is generally recognized, it is not always clear to what extent this principle is strictly adhered to. For instance, the psychiatric assessment is a multistage process involving a clinical interview supplemented by medical records and family history data, and a best-estimate diagnosis based on all sources of information. Each stage may involve several investigators. Ideally, blindness with respect to marker status should be maintained throughout the diagnostic process, but this is not always made explicit in published reports. Genetic marker analysis likewise involves several stages ranging from sample preparation to interpretation of autoradiograms. To ensure strict blindness, all individuals involved in the laboratory work must be blind to the clinical data. As for multiple analyses, investigators may be tempted to present the "best case" scenario, specifically, the genetic model and diagnostic scheme that yield the highest lod score. Other analyses with less favorable results may go unreported. The temptation to discard data which would not be compatible with linkage might lead to a similar outcome. The solution to these potential problems is to provide a thorough description of methods and data analysis. This would enable other investigators to interpret the results in light of the aforementioned methodological concerns.

Examples

The potential biases in the recent, well-publicized genetic findings have been examined. As noted earlier, several such reports have appeared. Linkage has been claimed for bipolar affective disorder and marker loci on chromosome 11p 13-15 (Egeland et al. 1987), Xq28 (Baron et al. 1987), and Xq27 (Mendlewicz et al. 1987).

All three studies appear to have utilized blind procedures for data collection and analysis. However, there were differences in the degree to which blindness was maintained. In the Egeland et al. (1987) study, DNA samples were typed blind to diagnosis. Best estimate diagnoses were made by a psychiatric panel blind to patient identity (presumably the marker status was also kept unknown although this was not made explicit). Unfortunately, it is not at all clear whether the raw clinical data were gathered blind to marker status given that the psychiatric status was periodically updated in this population. In the Mendlewicz et al. (1987) report, psychiatric interviews and diagnosis were carried out blind to the laboratory results but there was no indication that marker status was determined blind to the clinical data.

One of these studies (Egeland et al. 1987) involved multiple analyses of various model parameters and phenotype definitions. As pointed out by Edwards and Watts (1989), this problem was compounded by the large number of markers analyzed in an earlier study of the same population (Kidd et al. 1984). The maximum lod score in this study was 4.9. A deflation of the lod score to account for these multiple analyses would likely result in modest, inconclusive lod values.

When ascertainment is considered, studies involving series of pedigrees (Baron et al. 1987; Mendlewicz et al. 1987) reportedly were based on systematic screening of patient populations. (The Egeland et al. sample consisted of a single pedigree.) The maximum lod score in the Baron et al. study was largely due to high individual lod scores in three of the five reported pedigrees. The data overall gave evidence of linkage heterogeneity. By contrast, 8 of the 11 pedigrees reported by Mendlewicz et al. had small positive lod scores which, when combined, yielded a cumulative score that was interpreted as evidence of linkage; the sample appeared homogeneous with respect to linkage. (The three "negative" pedigrees had very small lod scores.) There is no readily available explanation for the nearly homogeneous linkage data in the Mendlewicz et al. series, which was based on a general clinical population. It stands to reason that a less uniform pattern of results should have been expected given the likely heterogeneity of affective illness.

Finally, one of the three positive studies yielded a borderline lod score. The maximum lod value reported by Mendlewicz et al. (1987) was 3.1 at complete penetrance; at 80% penetrance (probably a more realistic value), the lod score dropped to 2.4. When allowance was made for misclassified cases, the evidence for linkage all but vanished (Suarez et al. 1987). In the light of these issues, the failed attempts to reproduce most of these findings may not be surprising.

Type II Errors

Some circumstances that lead to type I errors can also generate false-negative results. Specifically, in the event of a bias *against* a particular linkage hypothesis, non-blindness and selective presentation of data will result in type II errors. By and large, however, type II errors may occur due to the genetic and phenotypic complexities of the disorders being investigated. These include etiologic heterogeneity, unknown mode of inheritance, phenocopies, assortative mating, and cohort effect.

1. Etiologic heterogeneity. If the fraction of linked cases is small (i.e., $\leq 20\%$), the sample required to detect linkage to a particular locus will have to be much larger than what is considered realistic under the usual sampling scheme.

2. Mode of inheritance. In some instances, misspecification of genetic parameters (penetrance, allele frequency), a likely occurrence when the underlying model is not known, may reduce the expected lod score (Clerget-Darpoux et al. 1986). This

is less likely to happen when a range of models is examined; the trade-off, however, is that multiple analysis may artificially inflate the lod score (see "Type I Errors"). A sharper drop in the lod score can occur with multipoint (as compared to pairwise) analysis due to model misspecification when a putative disease locus is placed between closely linked flanked markers. In this case, the use of high disease allele frequency may protect against a type II error (Risch and Giuffra 1990). Lod scores can also be deflated when the data are analyzed under the assumption of a single-gene model rather than several epistatically interacting loci, a likely genetic mechanism for some psychiatric disorders. It has been suggested, however, that if the correct model involves two loci, the assumption of a single locus may not lead to a substantial reduction in lod scores (Greenberg 1990).

3. Phenocopies. Misclassified cases, also known as phenocopies or false positives, increase the probability of type II errors. Phenocopies are likely present among the relatively common milder affective conditions which are often considered related to the rare, severe "core" phenotype. Restricting the disease phenotype to the severe end of the clinical spectrum and selection of pedigrees with high illness density (using restrictive diagnostic definitions) would curtail the number of phenocopies. Once linkage is detected with predetermined criteria, it would be of interest to examine other diagnostic schemes to see if they enhance or diminish the evidence of linkage. Should one opt for broad diagnostic boundaries, estimates of diagnostic uncertainty based, for example, on the stability and severity of the clinical phenotype could be factored in the analysis in order to minimize diagnostic misclassifications (Baron et al. 1990b).

4. Assortative mating. The impact of bilateral or bilineal transmission on linkage analysis is well known. Affected spouses would augment genetic heterogeneity within the pedigree with a resultant increase in phenocopies and, consequently, attenuated evidence of linkage. Assortative mating is ubiquitous in some psychiatric disorders, especially affective illness (Merikangas and Spiker 1982). To rule out bilineal inheritance, spouses and their families need to be studied.

5. Cohort effect. A change in the rate of affective disorder over time, or across generations, is well documented for (Klerman et al. 1985; Gershon et al. 1987). The underlying mechanism is not known; interaction between genetic disposition and environmental factors has been offered as one possible explanation. In linkage analysis, cohort effect is akin to a change in penetrance over time. To account for this phenomenon, the penetrance can be made to depend on cohort or year of birth. Failure to introduce a cohort adjustment could result in misspecification of genetic parameters and therefore a decrease in the expected lod score.

6. Laboratory errors. Random errors in reading marker alleles can lead to massive reductions in the maximum expected lod score (in some cases up to 50%) and overestimation of the recombination fraction (Terwilliger et al. 1990). Quality control of laboratory data, including careful labeling of tubes to avert sample mixups, duplicate typings, DNA fingerprinting to verify biological kinship, scoring genotypes by independent raters, double checking of unusual crossovers and, eventually, automation, could provide safeguards against such errors.

Examples

The failed attempts to reproduce some of the reported linkages in affective disorders deserve scrutiny. There have been six such reports on chromosome 11p 13-15 (Hodkinson et al. 1987; Detera-Wadleigh et al. 1987; Gerhard et al. 1988; Gill et al. 1988; Wesner et al. 1990; Mitchell et al. 1991), one report on chromosome Xq28 (Berrettini et al. 1990), and one on chromosome Xq27 (Gejman et al. 1990). The following issues are examined.

Blindness. Interestingly, no mention of blindness regarding clinical and genotypic status was made in five of the six chromosome 11p studies. Partial blindness -- with respect to genotypic (but not phenotypic) assignment — was noted in one of the other studies (Mitchell et al. 1991). A similar incomplete blindness was apparently present in the two X-chromosome studies (Berrettini et al. 1990; Gejman et al. 1990). Since blind procedures for genotypic and phenotypic assignments are such an obvious methodological issue, the failure to note blindness might have been an oversight. This, however, requires clarification.

Selection Bias. As with the positive reports, it is hard to assess whether there has been a preferential selection of data. However, in one of the cited reports this may have occurred. In the failed attempt to reproduce Xq28 linkage in bipolar affective illness, Berrettini et al. (1990) included a pedigree with clear male-to-male transmission. Not surprisingly, this pedigree yielded a negative lod score (for further discussion see Baron 1991b).

Etiologic Heterogeneity. Assuming a rare disease gene, none of the negative reports included a large enough number of pedigrees to rule out with certainty a type II error. The number of pedigrees in the various studies ranged from 1 to 15, reportedly with no evidence for linkage heterogeneity. In all, there were 11 pedigrees in the chromosome 11p studies, 15 in the Xq28 study, and 7 in the F9 study. However, owing to diagnostic uncertainty and assortative mating (see below), the number of pedigrees meeting rigorous criteria for linkage analysis is likely much smaller. For example, only 4 of the 15 pedigrees claimed by Berrettini et al. (1990) to exclude Xq28 linkage might qualify for X-linkage analysis (Baron 1991b).

Mode of Inheritance. The various studies used a realistic range of genetic parameters. Most of these studies employed several models in an attempts to capture the true genetic configuration. However, none of the multipoint analyses (Detera-Wadleigh et al. 1987; Hodkinson et al. 1987; Berrettini et al. 1990; Gejman et al. 1990) utilized disease allele frequencies thought to offset deflated lod scores (Risch and Giuffra 1990). Whether or not this has contributed to negative lod values in some of these studies is unclear.

Phenocopies. The extent to which phenotypic variation was taken into account varies greatly among the different studies. Three of the chromosome 11p 13-15 studies (Hodkinson et al. 1987; Gerhard et al. 1988; Wesner et al. 1990) employed a broad definition of affective illness, including both bipolar and unipolar disorders as affected. The other three studies (Detera-Wadleigh et al. 1987; Gill et al. 1988; Mitchell et al. 1991) utilized a narrower disease phenotype consisting of bipolar illness (Detera-Wadleigh et al. and Mitchell et al. used broader diagnostic boundaries as well; the diagnoses in the data of Gill et al. were largely of the bipolar type). However, only Mitchell et al. made a distinction in their analysis between bipolar I and bipolar II disorders in any of these phenotypic uncertainty studies. The same applies to the Xq28 and Xq27 studies (Berrettini et al. 1990; Gejman et al. 1990); the Xq28 study, in addition, included several pedigrees with low illness density. It is likely that the clinical diversity in the various pedigrees generated lod scores in some of the negative studies which were lower than expected.

Assortative Mating. Only three of the various reports on affective disorders (Hodkinson et al. 1987; Wesner et al. 1990; Mitchell et al. 1991) stated that they established firmly unilineal transmission in their pedigrees. Gill et al. (1988) noted that unilateral inheritance was likely but could not be proven. Detera-Wadleigh et al. (1987) and Berrettini et al. (1990) made no mention of measures to exclude assortative mating. In fact, Berrettini et al. appeared to have included in their sample pedigrees with bilineal transmission, as did Gejman et al. (1990) who studied some of the same pedigrees (Baron 1991b).

Cohort Effect. Systematic epidemiological data will be required to fully appreciate the extent of cohort effect in the various samples under consideration. Not surprisingly, an adjustment for cohort effect in a recent reanalysis of X-linkage data in bipolar affective disorder yielded increased lod scores (Van Eerdewegh 1989).

Laboratory Errors. The extent to which random laboratory errors have contributed to deflated lod scores is unclear. None of the negative studies made specific mention of scoring genotypes by independent raters. The other precautions noted earliest (e.g., duplicate typings) were not mentioned in any of the reports.

It cannot be overemphasized that negative linkage data should be subjected to the same rigorous scrutiny as positive findings owing to the multiple possible sources of type II errors (Baron 1991a). In the long term, false-negative linkage may prove more damaging than false-positive results because it may lead to the premature exclusion of genomic regions which, in fact, contain mappable disease genes.

Lessons To Be Learned

Several steps can be taken to avert spurious linkage results and to further the task at hand.

1. An emphasis can be placed on a narrowly defined disease phenotype, thus averting diagnostic uncertainties which are more likely to occur under broader diagnostic criteria. Although phenotypic variation can be partly due to pleiotropy (i.e., multiple phenotypic effects produced by a single mutant gene), at the present state of knowledge a conservative approach to the definition of the phenotype will likely reduce diagnostic error.

2. To further increase diagnostic accuracy (especially when the milder phenotypes are concerned), the probability of a case being true can be made to be a function of illness severity (i.e., number of symptoms, age at onset, recurrence, chronicity and incapacitation, response to treatment), diagnostic stability (as judged, for example, by longitudinal follow-up), and the amount and quality of the clinical material upon

which the diagnosis is based. To an extent this is analogous to converting a dichotomous trait (an all-or-none diagnosis) to a quasi-continuous one (Rice et al. 1987; Baron et al. 1990b).

3. To obviate the question of false negatives, more credence can be lent to an analysis restricted to affected individuals (the equivalent of a penetrance-free model in the lod score approach, or, when appropriate, the affected sib-pair analysis).

4. To improve diagnostic reliability and to avert diagnostic drift, much attention should be paid to training, repeated observations, and quality control both within and between centers engaged in these studies.

5. A search for clinical and biological correlates should be conducted to sort out homogeneous disease categories which may prove more amenable to the linkage approach than the larger universe of patients and families. Information on several traits that are affected by the same major gene may augment power to detect linkage. This approach could also serve to refine the clinical diagnosis and to attain further reduction in diagnostic error. In the absence of prior knowledge of distinct homogeneous illness types, linkage data can serve to define such disease categories. Specifically, discrete genetic forms identified by linkage analysis can be studied for unique phenotypic characteristics that set them apart from other genetic or nongenetic forms of the illness.

6. To increase homogeneity within families, the availability of several informative relatives with a narrowly defined phenotype should be considered an inclusion criterion. Families, or branches thereof, with bilineal transmission should be excluded pending further development in analytical methodology.

7. To eliminate a potential source of systematic bias, diagnostic determinations and marker typing should be carried out blind to each other.

8. To minimize the risk of false linkage, phenotype definitions and model parameters should be selected *prior* to the analysis. Combined segregation and linkage analysis can also be performed, although successful inferences from this analysis can be made only for extended data sets (e.g., pedigree data). The number of permutations should be limited, and, when necessary, the linkage statistic should be adjusted for multiple test effects. To increase confidence in the results, the lod criterion may have to be set higher than the conventional value of 3.

9. Careful laboratory work should be carried out along the lines proposed earlier which is aimed at minimizing random errors in the assignment of genotypes, a scenario associated with deflated lod scores.

10. Replication is the ultimate criterion by which confidence in a given finding is established. However, in the near term, replication may not be an easy task to accomplish due to etiologic heterogeneity and other complicating factors associated with type II errors. To reinforce (or refute) a given finding, the following interim strategies can be employed: (a) Extension of pedigrees with reported linkages to determine whether the same segregation pattern exists in other branches of the family. (b) Updating and periodic reevaluation of the clinical material as a safeguard against diagnostic misclassifications. (c) Typing new markers in the target genomic region to confirm and refine the location of the putative gene. As noted earlier, the extension and updating of clinical information has led to the demise of the

chromosome 11p linkage in the Amish study (Kelsoe et al. 1989). A reanalysis of the Xq28 data, using a DNA marker in the vicinity of F9, has shown that the new locus, DXS98, is not linked to the illness (Freimer et al. 1989). This would not be compatible with the suggestion made by Mendlewicz et al. (1987) that their F9 data support the earlier reports on Xq28 linkage in bipolar affective illness. (d) For a large enough sample, a half-sample approach can be employed whereby half the data are used for selecting the most plausible model, whereas the second half of the data is analyzed under the chosen fixed model (Weeks et al. 1990). This strategy offers a measure of cross validation for linkage results.

11. As I noted earlier, mechanisms underlying the major psychoses may include major genes, multiple loci and interactions, and nongenetic etiologies. Predictions concerning the prevalence of the various disease forms are premature, but the failure of segregation analysis to demonstrate the general presence of major gene effects raises the possibility that discrete single genes account for only a minority of cases. In any event, to prove linkage heterogeneity, it will be necessary to demonstrate linkage to at least two distinct loci. To be sure, even when linkage to a particular locus is proven, possible contributions from other genes (presumably of minor effect) to genetic susceptibility in the *same* pedigrees should be evaluated.

12. The successful application of genetic linkage strategies can, in principle, cast light on gene–environment interaction. For example, comparisons can be made between affected and older unaffected family members (assuming the latter have completed their risk period) who are identical for the marker locus to determine whether particular environmental factors are more likely to be present in either group of subjects. Similarly, prospective studies of at-risk subjects identified by a linked marker could reveal specific environmental influences which enhance or suppress the expression of illness.

13. Because genes for mental illness may give rise to a broad spectrum of behavioral manifestations (pleiotropic effects) ranging from psychopathology to normal variation of behavior, the elucidation of genetic factors in etiology could ultimately lead to significant insights into other psychiatric traits and human behavior in general. For example, once linkage is established, potential phenotypic variants that represent broader definitions of the disease phenotype or fall outside the conventional diagnosis altogether (personality traits) can be explored with respect to their genetic relatedness to the core phenotype.

14. Further advance in molecular and quantitative genetics is required in several areas: a dense linkage map of polymorphic markers equally spaced along the human genome, allowing a total search for putative disease loci; the identification of rare disease genes when large amounts of genetic material (i.e., informative meiotic events) are not available; the detection of multiple interacting genes, as well as genes with common alleles, that account for small amounts of phenotypic variation; models invoking continuous, rather than dichotomous (all-or-none) phenotypes, such as behavioral dimensions and biological endophenotypes; and methods incorporating several traits (qualitative as well as quantitative) in the linkage analysis in the event these traits co-vary with the illness and may thus represent alternate expression of the same major gene. New paradigms involving sampling

strategies, analytical design, and new statistical methods appropriate for the complex nature of the data being analyzed are called for to address these issues.

Conclusion

On close scrutiny, some of the linkages hitherto reported may be chance findings. The remainder are not without uncertainties; for the time being, they should be best considered as hypotheses. The many complicating factors in the genetic analysis of mental disorders, including crude diagnostic assessment, phenotypic diversity, etiologic heterogeneity, the number of loci involved and their interactions, misspecified genetic parameters, and variable age at onset, impose limitations on the ability of current models to tease out specific genetic mechanisms. That is not to say that affective disorders are genetically intractable. With improved methodology, suitable data, and joint effort by clinical researchers, statistical geneticists, and molecular biologists, genetic factors in etiology will come to light.

References

American Psychiatric Association (1987) Diagnostic and statistical manual of mental disorders (DSM-III-R). APA, Washington, DC

Baron M (1990a) Genetic linkage in mental illness. Nature 346:618

Baron M (1990b) Molecular genetic studies in affective disorders. In: Bunney WE Jr, Hippius HH, Laakmann G, Schmauss M (eds) Neuopsychopharmacology. Springer, Berlin Heidelberg New York, pp 108-116

Baron M (1991a) Genetic mapping of affective disorders: opportunities and pitfalls. In: Racagni G et al (eds) Biological psychiatry 1991. Elsevier, Amsterdam. (In Press)

Baron M (1991b) Genetics of manic depressive illness: current status and evolving concepts. In: McHugh PR, McKusick V (eds) Genes, brain and behavior. Raven, New York, pp 153 161

Baron M (1991c) X-linkage and manic-depressive illness: a reassessment. Soc Biol (In press)

Baron M, Rainer JD (1988) Molecular genetics and human disease. Implications for modern psychiatric research and practice. Br J Psychiatry 152:741-753

Baron M, Risch N, Hamburger R et al. (1987) Genetic linkage between X-chromosome markers and bipolar affective illness. Nature 326:289-292

Baron M, Endicott J, Ott J (1990a) Genetic linkage in mental illness. Limitations and prospects. Br J Psychiatry 157:645-655

Baron M, Hamburger A, Sandkuyl LA et al. (1990b) The impact of phenotypic variation on genetic analysis: application to X-linkage in manic-depressive illness. Acta Psychiatr Scand 82:196-203

Berrettini WH, Goldin LR, Gelernter J, Gejman P, Gershon ES, Detera-Wadleigh S (1990) X-chromosome markers and manic-depressive illness. Arch Gen Psychiatry 47:366-373

Bishop DT, Williamson JA (1990) The power of identity-by-state method for linkage analysis. Am J Hum Genet 46:254-265

Buckle VJ, Fujita N, Ryder-Cook AS et al. (1989) Chromosomal localization of GABA receptor subunit genes: relationship to human genetic disease. Neuron 3:647-654

Cavalli-Sforza LL, King M-C (1986) Detecting linkage for genetically heterogeneous diseases and detecting heterogeneity with linkage data. Am J Hum Genet 38:599-616

Clerget-Darpoux F, Bonaiti-Pellie C, Hochez J (1986) Effects of misspecifying genetic parameters in lod score analysis. Biometrics 42:393-399

Clerget-Darpoux F, Babron M-C, Bonaiti-Pellie C (1990) Assessing the effect of multiple linkage tests in complex diseases. Genet Epidemiol 7:245-253

Crow T (1988) Sex chromosomes and psychosis: the case for a pseudoautosomal locus. Br J Psychiatry 153:675-683

Detera-Wadleigh SD, Berrettini WH, Goldin LR, Boorman D, Anderson S, Gershon ES (1987) Close linkage of c-Harvey-ras-1 and the insulin gene to affective disorder is ruled out in three North American pedigrees. Nature 325:806-808

Edwards J, Watts D (1989) Caution in locating the gene(s) for affective disorders linked to DNA markers on chromosome 11. Psychol Med 19:273-275

Egeland J, Gerhard D, Pauls DL et al. (1987) Bipolar affective disorder linked to DNA markers on chromosome 11. Nature 325:783-787

Elston RC, Wilson AF (1990) Genetic linkage and complex diseases: comment. Genet Epidemiol 7:17-20

Freimer N, Sandkuyl LA, Baron M et al. (1989) Genetic mapping of the Xq28 region using CEPH and bipolar pedigrees. Cytogenet Cell Genet 51:1000

Gejman PV, Detera-Wadleigh S, Martinez MM, Berrettini WH, Goldin LR, Gelernter J, Hsieh W-T, Gershon ES (1990) Manic depressive illness not linked to factor IX region in an independent series of pedigrees. Genomics 8:648-655

Gerhard DS, Todd R, Devor E, Reich T (1988) Linkage analysis of polymorphic markers on two chromosomes with affective disorders. Am J Hum Genet 43 [Suppl]: A144

Gershon ES (1990) Genetics. In: Goodwin FK, Jamison KR (eds). Manic-depressive illness. Oxford University Press, New York, pp 373-401

Gershon ES, Hamovit JH, Guroff JJ, Nurnberger JI (1987) Birth-cohort changes in manic and depressive disorders in relatives of bipolar and schizoaffective patients. Arch Gen Psychiatry 44:314-319

Gill M, McKeen P, Humphries P (1988) Linkage analysis of manic depression in an Irish family using H-ras I and INS DNA markers. J Med Genet 25:634-637

Goldin LR (1990) The increase in type I error rates in linkage studies when multiple analyses are carried out on the same data; a simulation study. Am J Hum Genet 47:A180 (0707)

Green P (1990) Genetic linkage and complex diseases: comment. Genet Epidemiol 7:25-28

Greenberg D (1990) Linkage analysis assuming a single-locus mode of inheritance for traits determined by two loci. Genet Epidemiol 7:467-479

Gurling HMD (1985) Application of molecular biology to mental illness. Analyses of genomic DNA and brain mRNA. Psychiatr Devel 3:257-273

Hodkinson S, Sherrington R, Gurling H, Marchbanks R, Reeders S, Mallet J, McInnis M, Petursson H, Brynjolfsson J (1987) Molecular genetic evidence for heterogeneity in manic depression. Nature 325:805-806

Holland T, Gosden C (1990) A balanced chromosomal translocation partially co-segregation with psychotic illness in a family. Psychiatry Res 32:1-8

Holmes D, Brynjolfsson J, Brett P et al. (1991) No evidence for a susceptibility locus predisposing to manic depression in the region of the dopamine (D2) receptor gene. Br J Psychiatry 158:635-641

Kelsoe JR, Gings EI, Egeland JA et al. (1989) Reevaluation of the linkage relationship between chromosome 11p loci and the gene for bipolar affective disorder in the Old Order Amish. Nature 342:238-243

Kidd KK, Gerhard DS, Kidd JR, Housman D, Egeland JA (1984) Recombinant DNA methods in genetic studies of affective disorders. Clin Neuropharmacol 7[Suppl 1]:5104

Klerman GL, Lavori PW, Rice J et al. (1985) Birth cohort trends in rates of major depressive disorders among relatives of patients with affective disorders. Arch Gen Psychiatry 42:689-695

Lander E, Botstein D (1986) Strategies for studying heterogeneous genetic traits in humans by using a linkage map of restriction fragment length polymorphisms. Proc Natl Acad Sci USA 83:7353-7357

Leboyer M, Babron MC, Clerget-Darpoux F (1990) Sampling strategy in linkage studies of affective disorders. Psychol Med 20:573-579

Majumder PP (1989) Strategies and sample size considerations for mapping a two-locus autosomal recessive disorder. Am J Hum Genet 45:412-423

Martinez M, Goldin LR (1989a) The detection of linkage and heterogeneity in nuclear families for complex disorders: one versus two marker loci. Am J Hum Genet 44:552-559

Martinez M, Goldin LR (1989b) Systematic genome screening for heterogeneous disorders using map intervals. Am J Hum Genet 47:A1901 (0749)

Mendlewicz J, Simon P, Sevy S et al. (1987) Polymorphic DNA marker on X-chromosome and manic depression. Lancet i:123-1232

Merikangas KR, Spiker DG (1982) Assortative mating among inpatients with primary affective disorder. Psychol Med 2:753-764

Mitchell P, Waters B, Morrison N et al (1991) Close linkage of bipolar disorder to chromosome 11 markers is excluded in two large Australian pedigrees. J Affect Disorder 21:23-32
Ott J (1986) The number of families required to detect or exclude linkage heterogeneity. Am J Hum Genet 39:159-165
Ott J (1989) Computer-simulation methods in human linkage analysis. Proc Natl Acad Sci USA 86:4175-4178
Plomin N (1990) The role of inheritance in behavior. Science 248:183-188
Rice JP, Endicott J, Knesevich MA et al. (1987) The estimation of diagnostic sensitivity using stability data: an application to major depressive disorders. J Psychiatr Res 21:337-345
Risch N (1990a) Genetic linkage and complex diseases, with special reference to psychiatric disorders. Genet Epidemiol 7:3-16
Risch N (1990b) Linkage strategies for genetically complex traits, I. Multilocus models. Am J Hum Genet 46:222-228
Risch N (1990c) Linkage strategies for genetically complex traits, II. The power of affected relative pairs. Am J Hum Genet 46:229-241
Risch N, Giuffra L (1990) Multipoint linkage analysis of genetically complex traits Am J Hum Genet 47: A197(0772)
Robb AS, Berrettini WH, Detera-Wadleigh SD et al. (1991) Linkage markers of the pseudoautosomal region in bipolar disorder. Biol Psychiatry 29:338S
St. Clair D, Blackwood D, Muir W et al. (1990) Association within a family of a balanced autosomal translocation with major mental illness. Lancet 339:13-16
Sandkuyl LA, Ott J (1989) Determining informativity of marker typing for genetic counseling in a pedigree. Hum Genet 82:159-162
Smith M, Wasmuth J, McPherson JD et al. (1989) Co-segregation of an 11q22.3-9p22 translocation with affective disorder; proximity of the dopamine D-2 receptor gene relative to the translocation breakpoint. Am J Hum Genet 4:A220
Spitzer RL, Endicott J, Robins E (1978) Research diagnostic criteria: rationale and reliability. Arch Gen Psychiatry 35:773-779
Suarez BK, Hampe CL, Wright AF (1987) Linkage analysis in manic-depressive illness. Lancet ii:345-346
Suarez BK, Reich T, Rice JP et al. (1990) Genetic linkage and complex diseases: comment. Genet Epidemiol 7:37-40
Terwilliger JD, Weeks DE, Ott J (1990) Laboratory errors in the reading of marker alleles cause massive reductions in lod score and lead to gross overestimates of the recombination fraction. Am J Hum Genet 47:A201 (0790)
Thompson EA (1984) Interpretation of LOD scores with a set of markers loci. Genet Epidemiol 1:357-362
Van Broeckhoven C, Mendlewicz J (1991) Molecular genetic analysis in bipolar illness. Biol Psychiatry 29:12S
Van Eerdewegh P (1989) Linkage analysis with cohort effect: an application to X-linkage. Genet Epidemiol 6:271-276
Weeks DE, Lehner T, Squires-Wheeler E et al. (1990) Measuring the inflation of the lod score due its maximization over model parameter values in human linkage analysis. Genet Epidemiol 7:237-244
Wesner RB, Scheffner W, Palmer PJ et al. (1990) The effect of comorbidity and penetrance on the outcome of linkage analysis in a bipolar family. Biol Psychiatry 27:241-244
Zimmerman M (1988) Why are we rushing to publish DSM-IV? Arch Gen Psychiatry 45:1135-1138

Discussion of Presentation M. Baron

Mendlewicz

Since you have cited some of our work on the factor IX anylsis in manic depression, may I add something. The overall lod scores we found in a group of 10 small size families have never been very robust. Furthermore, if we used a variable penetrance value, which is more reasonable than full penetrance, the lod score fell below the value of 2. So indeed, this is not a very robust lod score, especially in the area of psychiatric disorders and we all agree on the necessity of much higher lod scores to be convinced of the reality of the results. There is one question I would like to ask you. In one of your slides on the data on the X-chromosome you did not include your own positive results on color blindness. Since you have mentioned that you are doing a follow-up study on the original pedigrees, do you have any results using DNA markers on these pedigrees?

Baron

We are actively pursuing the various aspects of this particular work, including extension of the pedigrees, updating diagnoses, and typing every known DNA marker in the Xq27-28 region. We do not have the final results yet. The only result we do have was presented a couple of years ago by Freimer et al. from our group (First World Congress on Psychiatric Genetics). In that study, DXS98, a DNA marker in the Xq27 region, did not show linkage to bipolar illness. In other words, there is either a second disease locus in the factor IX area in some pedigrees or else the factor IX linkage you and colleagues have reported was a chance finding. As far as the Xq28 region is concerned, we do not have final results.

Barnard

I would like to point to the fact that the linkage program is so heavily biassed, assuming a single genetic locus. It is really rare to have a single locus contributing to a psychiatric disease, even if you take the most apparently isolated populations, such as the Old Amish or Icelandics. It still seems to me to be an optimistic assumption that there will be one locus, when there may be several genetic groups to a psychiatric disorder. If you have two major components in the population, then of course the lod score must be weakened. Therefore, one should look for alternative statistical analyses, which we certainly have. Jacques Mallet recently presented at the World Congress on Biological Psychiatry that he had reanalyzed the Old Amish Data of Egeland and Kidd. He was including the new data without assuming a direct linkage and found in those families a good association with chromosome 11. So it seems to me that the data may be correct but the method of analysis may be wrong.

Baron
Several groups are considering methods for additional features in linkage analysis, including a number of different genes. However, if you were to assume a single gene in your computer program, whereas the real case scenario calls for a number of genes, you would probably not come up with false evidence for linkage. In fact, it may be the other way around, that is, obtaining deflated lod scores. We have some computer experts here. Jürg Ott, for example, may wish to comment on this issue.

Ott
I do not think we should now damn the statistical programs. The programs are designed to do a specific task and when they are used according to their original design, then they are fine. Now, you were speaking of different genes. It depends a little on whether one assumes that you also could have different genes in the same family — or whether you assume that in each family there is only one gene. But, in the population you will have different genes. In this case, when you have different genes in the population, but only one in each family, then there are well-known ways of analyzing the data. One could use these programs the way they are and apply tests for heterogeneity afterwards on the basis of the lod scores. But there have also been developments by which one can analyze data under the assumption that there may be two interacting genes jointly causing disease. This is also an extension of the linkage programs that we presented a year ago at the Annual Meeting of the American Society of Human Genetics. I think the technical facilities are present.

Baron
I would like to make one comment. I mentioned tyrosine hydroxylase earlier. I think the results are quite mixed, and other groups have not been able to replicate any association with tyrosine hydroxylase. Whether that is the case in the Amish population, I don't know, but it is not that clear-cut.

Kidd
I will just add a couple of comments on the Amish study. First, one of the original criticisms about multiple tests is not quite appropriate and applicable, because the chromosome 11 findings were made on the very first DNA markers that were studied in a pedigree. The previous linkage studies had been done in the Amish on an unknown identical, but partly overlapping, set of family material. And most of those classical genetic markers were not informative. So there were really no prior informative lod scores that one could use to correct and come up with an estimate. We now do have in press a paper in *Human Genetics*, reporting on 190 informative loci in the Old Order Amish pedigree. And we can exclude 25% of the genome, this done with all of the caveats of the sort you just mentioned in your excellent summary. But in that set, the highest lod scores that exist for the set of markers, typed only in the original family, are still HRAS and insulin with lod scores slightly greater than

2. The other interesting development is that we have just completed the linkage mapping of DRD4 maps right next to the HRAS locus.

Baron

This is a very interesting comment. I know that some people are discouraged by the chromosome 11 data. Would you then recommend pursuing this?

Kidd

I really have to say, there is not significant evidence for a gene on chromosome 11. And my own basic philosophy is that it is a false positive but not a real phenomenon. But, the positive lod scores still exist for a region that includes tyrosine hydroxylase and the D_4-dopamine receptor in that original section of the Amish pedigree. Now, the posthoc argument to say that the extension may involve a different locus segregating does not illuminate the overall extremely negative lod scores. This is a very real problem that we have to deal with, knowing when to be conservative and assume homogeneity and a real phenomenon existing, or when to back off and say there may be heterogeneity and in this subsection of the data there may be such a locus. It is not a question that conventional statistics readily deals with. It is a matter of what the real situation is and unfortunately we don't know what the real situation is yet.

Gurling

Just two points of information. One refers to Eric Barnard's comment about the reanalysis of the Amish data carried out by Jacques Mallet. In fact the analysis consisted of using the sib-pair method and apparently there is a significant result using tyrosine hydroxylase polymorphisms in the Amish data. Secondly, Jacques Mallet increased the sample size of bipolar individuals, unrelated individuals, and so has seen an association. So we should bear in mind the possibility that there is some epistatic effect from tyrosine hydroxylase. It is, after all, on the dopamine pathway, so that it is possible that in parts of the pedigrees it might simulate major locus effect, but in general simply modulates or affects the outcome for someone who has inherited a gene of another locus.

Kidd

My statement that the positive lod scores still exist in that first set of families could very well explain the association studies, since that is about three quarters of the total sample. It does not explain the unrelated control association with tyrosine hydroxylase.

Reich

The current viable model with respect to the minor allele of the tyrosine hydroxylase, although I think there are several negative as well as positive studies, is that the manic-depressive gene is located somewhere else, but the minor allele for tyrosine hydroxylase is under influence of penetrance. Accordingly, there is a disequilibrium with respect to bipolar illness. I wanted to ask Dr. Ott if that was the situation, namely, that you had two loci, epistatically interacting, one of which was in equilibrium and the other one in disequilibrium. Do you think that would yield readily or produce the situation we see, where the lod score is very largely negative, minus 5, perhaps more, but there could be a positive significant association with a common allele at that locus with the disease? So the two questions of you are: do you think it is possible that that could occur and, secondly, using the two locus linkage programs, could one analyze the data with that in mind?

Ott

I am not really in a position to decide whether it could occur, but I am sure it could be that two locus linkage programs could be used exactly in that situation. I have actually discussed this problem with Janice Egeland. They are planning on using the two locus models on their families. Thus this would be a very good data set for these programs.

McGuffin

We should certainly not last too much to the discussion of tyrosine hydroxylase, but to something that Ken Kidd said a little earlier. In their studies, they have obviously excluded a gene for manic depression in about 25% of the genome. But I assume he means that he excluded a gene from 25% of the genome, assuming a particular model, and I assume that they are assuming a near double model. Is that right?

Kidd

Yes, as I said, when I made that statement with all the usual caveat. We have excluded that particular model; that is the best one supported by pedigree analysis in the Amish and in fact was the model used by Kelsoe and colleagues.

Hebebrand

Would you also consider formal genetics as a further pitfall? In your *Nature* paper you have several affected men, but only five of them have children. One married into the pedigree and four have children that are too young to have reached the manifestation age. Now, in your linkage studies you assume of course an X-linked dominant mode of inheritance and for that you would require that all the daughters of affected men should be of high penetrance rate, which indeed has been so. If you look at all the X-linkage studies, you actually do not find many instances of males who have reproduced. Mostly you cannot see the characteristic segregation ratio you do expect in X-linked dominant disorder.

Baron

I have not examined carefully other X-linkage studies from this vantage point. Since much of the X-linkage data was provided by Dr. Mendlewicz, perhaps he would like to comment on this issue. As far as our particular research is concerned, it is part of our follow-up study to look at male to male transmission, that is, the clinical status of male offspring of affected males. This, indeed, is another feature that complicates and lengthens our follow-up study. Some of our affected males have reproduced and some have not. Much has been said about reduced reproduction in schizophrenia, but there is also reduced reproduction in bipolar illness. And there are a number of interesting reports on this phenomenon. In our experience there in an increased rate of affected males who have not reproduced.

Gurling

There is a study published in the *Australian and New Zealand Journal of Psychiatry* reporting on pedigrees of manic depression and color blindness cosegregating. The authors did not calculate a lod score, but there are other references which generally are not quoted in the literature, which tend to support the X-linkage. This is one that I think people should have another look at.

Maier

When using the linkage tests we have three decisions: that there is linkage, that there is no linkage, that we have no decision and we have to proceed. Now, the list of type 2 errors, as far as I understood it, was related to the decision that you are accepting linkage. Are they also valid for the decision that you are excluding linkage?

Baron

The caution I advised earlier has to do with a situation in which you have a study reporting a negative finding, presumably a nonreplication of a positive finding, and you have to ask yourself to what extent that conclusion (that is, the purported nonreplication) is really valid or can be fully accepted. That is what I meant in my list of potential type 2 errors.

Crow

Can I take issue with your comments about bilineality and the mode of transmission. You made a recommendation that you should exclude such families, a general recommendation, and you should assume a dominant mode of transmission. That may be plausible for the Amish pedigrees, but even there it seems that there are questions and we have to consider the possibility that we are actually dealing with very common genes that are additive and in some sense recessive. In that case you might make the contrary recommendation, namely, that you should adopt a sib-pair technique and make no assumption about the mode of transmission. But, I would suggest that should be a parallel strategy.

Baron

I think I was quite cautious in my approach to this issue. Assortative mating or bilineal transmission is a very interesting phenomenon, but we need more theoretical work, as to how one could incorporate this in the analysis for the sake of a better understanding of linkage results. I am not suggesting that all these pedigrees should be thrown into the wastebasket. We do retain many families with assortative mating for future reference and study. There may, indeed, be other paradigms to consider. The mode of transmission may be different. One could of course use the sib-pair method. But in terms of the particular approach that has been favored so far, and which certainly pertains to most of the published linkage studies, the assumption made is that you have a single gene. When you look at many of those pedigrees, they seem consistent with this assumption, namely, a gene predisposing to a dominant or nearly dominant trait rather than recessive transmission. So my talk may be perceived as somewhat biassed in this respect. But I do not exclude alternatives.

Van Broeckhoven

I would like to make some comments on the latest X-linkage studies which were recently presented. The study included 10 families, all selected from a larger set of families because they did not show any male to male transmission. But as a matter of fact, in one family an affected male did have an affected offspring. In that sample of 10 families, two families of the original study were included and these two families were still positive for factor IX. One of the families is negative for ST14; the other family was not informative. We have also done a multipoint study using factor IX and ST14, including a number of markers located in between factor IX and ST14. The overall lod scores are negative. But, if we look at the families separately and especially under the bipolar model, there are still some families that are not excluded from that region. So one should keep in mind that we have only presented the total lod scores for these 10 families and these lod scores are negative, but it does not go for each family included in the study.

Baron

Have you been able to do the typing you just mentioned on the previous series of families published in *Lancet* in 1987, those that were positive for factor IX?

Van Broeckhoven

The other families of the previous study have not been resampled because they were cooperative. We are still trying to persuade the families to work with us again since we would like to reanalyze all these pedigrees again.

Baron

It would have been of great interest for investigators with previously published positive results to go back to their pedigrees and to extend the clinical and genetic information. Especially the more impressive pedigrees, such as the Iranian pedigree you reported on by Dr. Mendlewicz and his colleagues in 1980.

Froster-Iskenius

It reminds me very much of what we found when we did the first linkage studies in fragile X families. This region of the X-chromosome seems to have a high recombination frequency. I remember that in the early days of doing linkage with fragile X families, we found linkage heterogeneity for families which were informative for factor IX and others which were not. So in fact, it is just the question of whether it is close enough.

Molecular Genetics of Schizophrenia

P. McGuffin, M. Owen and M. Gill

Introduction

Although the serious study of the genetics of psychiatric disorders is now a subject well into its seventh decade, never before has there been greater interest. This can largely be traced to the widespread optimism that molecular genetics will ultimately hold the key to our understanding of complex and common disorders such as schizophrenia, but is the current optimism justifiable? When it first became apparent that approaches from the "new genetics" might have an application in disorders such as schizophrenia, the reaction of some commentators was to point out the difficulties (Sturt and McGuffin 1985) or to advise cautiously that molecular genetics did not offer a panacea (Cloninger et al. 1983). These authors agreed that, although the techniques were of great interest, nobody should expect too much too soon in the way of tangible advances. Others regarded these views as hopelessly pessimistic and, encouraged by the advances in the understanding of single gene disorders, applied broadly similar approaches in their studies of the major psychoses. Soon apparently quite convincing positive linkage results emerged in studies of manic depression (Egeland et al. 1987) and schizophrenia (Sherrington et al. 1988) but neither of these provocative sets of findings proved capable of replication. In each case the response of interested observers was in turn to let out a gasp of delight followed by a groan of disappointment.

This was perhaps particularly true among those who had not previously taken any specialised interest in psychiatric genetics. Against this background there is a danger that molecular genetic studies of psychosis will be viewed as some temporarily fashionable blind alley from which it is time to make a hasty retreat. Therefore the aim of this paper is to strike a balance between overeager expectations on the one hand and dismal scepticism on the other. The main theme is that advances are likely to come from a methodical and systematic search through the human genome, probably requiring collaborative efforts from multiple centres, rather than from a rapid and dramatic breakthrough from one fortunate group of researchers on their own. We will restrict our attention purely to studies of linkage and association for, while we acknowledge that basic research holds great interest in such areas as gene expression (Harrison and Pearson 1989) or the genetic control of brain development (Bloom 1990), so-called reverse genetic approaches at present offer the most promising and practical ways of proceeding that are of direct relevance to the aetiology of schizophrenia.

The Prospects of Genetic Marker Studies

The two main strategies for investigating the relationship between genetic markers and disease are *association* and *linkage*. These are different but related. Association studies require no major assumption other than the probable existence of a genetic contribution to the disorder. The approach then is simply to investigate whether a particular marker allele is more common in a sample of schizophrenic individuals than in healthy controls. We will return to consider association studies later but for the most part we will concentrate on linkage studies of schizophrenia. These are more difficult to perform since they require the study of families rather than of unrelated individuals; however, it may be argued that they are easier to interpret than association studies and they certainly offer greater prospects for rapid advance since the aim is to *detect* genes of major effect. The process of detection depends on the demonstration of departure from independent assortment, in other words showing that the recombination fraction between the putative disease locus and a marker locus is less than a half.

The subsequent aim is then to *estimate* the size of the recombination fraction since, within certain limits, this is roughly proportional to the distance between the two loci. The next stage is to move from marker locus to the disease gene itself in a series of steps (Davies and Reid 1988) which include a search for more closely linked markers, and preferably markers which flank the disease gene. This process may be long and difficult. For example, the gene for Huntington's disease was localised to the short arm of chromosome 4 (Gusella et al. 1983) more than 8 years ago, but the disease gene itself is yet to be isolated. Nevertheless it is virtually certain that this or any other disease gene for which there is firm linkage data will be isolated, its DNA sequence studied and the gene products discovered. Hence reverse genetic strategies provide attractive prospects as a means of uncovering the biochemical "lesion" in a disorder when the pathogenesis was hitherto obscure, and thence offers prospects for the development of rational therapies. What then are the particular difficulties facing the investigator who wishes to carry out linkage studies in schizophrenia?

The Problems

The first obvious problem is one of diagnosis. It is widely believed among nonpsychiatric researchers that the greatest difficulty here is one of diagnostic unreliability. This is a misconception. Painstaking research over the past 20 years has resulted in methods of eliciting and recording symptoms and arriving at a definition of schizophrenia which affords excellent agreement between clinicians. However, the problem is that there is not one but many different operational definitions of schizophrenia available to the researcher, and although these look superficially similar they do not always match each other in practice (Farmer et al. 1991). The problem therefore is how to choose the definition which is most valid from the viewpoint of genetic research. Some guidance may be obtained from family or twin studies since a high level of heritability may be regarded as a prerequisite for the definition of a phenotype to be used in a linkage study. Unfortunately currently available data still does not allow us to choose between definitions. For example, the St. Louis criteria (Feighner et al. (1972), the Research Diagnostic Criteria (RDC), of Spitzer et al. (1978) and the DSM-III definition (American Psychiatric Association 1980) all result in a form of schizophrenia of high heritability (McGuffin et al. 1984; Farmer et al. 1987). However, when applied to the same series of psychotic patients each set of criteria defines a different (albeit overlapping) set of patients as having schizophrenia (Farmer et al. 1991). It would therefore seem advisable in linkage studies to perform the clinical assessment in a way which is both comprehensive and sufficiently flexible to allow more than one operational definition to be applied.

The difficulties about such a polydiagnostic approach is that it can soon become unwieldy and time-consuming and so, in the interests of maintaining utility, collaborative programmes such as the one recently established under the auspices of the European Science Foundation (ESF) combine the use of a research interview such as the Schedule for Affected Disorders, Lifetime version (SADS-L; Spitzer et al. 1978) or the 10th edition of the Present State Examination (PSE-10; Wing et al. 1990) with an operational criteria (OPCRIT) checklist (McGuffin et al. 1991). This provides a way of recording summary clinical material and can be applied when an interview is impossible but when hospital casenotes or other written material are available. Having completed the OPCRIT checklist a computer scoring program generates the classification according to a range of operational criteria.

A related issue is how to set boundaries on the phenotype. It is well established that among the biological relatives of schizophrenics there is an excess not just of schizophrenia but also of "fringe" phenotypes which, while not amounting to full blown schizophrenia, are recognisably different from normal (Gottesman 1987). These have in the past been referred to as spectrum disorders (Kety et al. 1976) but there is still considerable debate as to what can be legitimately included within the spectrum. There is good evidence that schizotypy, as comparatively narrowly defined by DSM-III, is genetically related to schizophrenia (Kendler et al. 1981) and there is suggestive evidence that atypical psychosis and affective disorder with mood incongruent psychotic symptoms should also be included within the spectrum (Farmer et al. 1987; Kendler et al. 1987). More controversial is how to classify relatives of schizophrenic patients who present with affective illness which is severe but lacking in psychotic features. Here, again, an exploratory approach is required in which linkage analysis is performed taking disease phenotypes of differing breadth. Clearly the decisions about breadth of definition, as with all other diagnostic decisions, need to be taken before seeing any marker data, and blindness must be preserved up until the stage of actually performing the linkage analysis.

How to embark on linkage analysis then presents the next major difficulty, given that the mood of transmission of schizophrenia is unknown. Pure single locus inheritance in which a major gene is the sole source of resemblance between relatives is unlikely (O'Rourke et al. 1982; McGue et al. 1985), but the possibility remains that there is a gene of major affect or several different major genes which, together with polygenic or multifactorial environmental factors, contribute to the transmission of schizophrenia. A related and unsettled issue concerns aetiological heterogeneity; thus it could be argued that even if schizophrenia as a whole is not a monogenic disorder, genes of major affect will be detectable in at least some families in which multiple members are affected over two or more generations. The difficulty then is knowing how to specify the values of the major gene parameters (i.e. the gene frequency and the penetrance vector). It is somewhat reassuring that errors in diagnostic classification or misspecification of the mode of transmission should not usually lead to spurious detection of linkage where in fact no linkage is present (Clerget-Darpoux et al. 1986). However, misspecification and misclassification can result in failure to detect linkage even when true linkage is present. Furthermore even if linkage is successfully detected applying an uncertain model disease transmission, the resulting estimate of recombination will also be uncertain. Thus the process of moving from marker locus to disease gene will inevitably be more difficult than for a disorder with a simple mendelian mode of transmission.

A related difficulty in the detection and estimation of linkage is that any plausible model of the transmission of schizophrenia involving a major gene must allow that the penetrances deviate from the mendelian values of 0 or 1. Penetrance and recombination are partially confounded since individuals who are unaffected but who are genotypically identical with their affective relatives at the disease locus may be classified as recombinants. Incomplete penetrance and age-related penetrance can be incorporated in the calculation of logarithm of odds (lod) scores and this is, for practical purposes, straightforward when using computer programs such as LINKAGE (Lathrop and Lalouel 1984) or LIPED (Ott 1985). However, the presence of incomplete penetrance reduces the efficiency of linkage analysis so that the sample size required for the confident detection of linkage is almost always much higher than if the main trait shows regular mendelian segregation with complete penetrance (Ott 1985; Sturt and McGuffin 1985).

Assumptions and Practical Procedures

Faced with these problems most researchers attempting linkage studies in schizophrenia have taken a pragmatic approach and have concentrated on 'loaded' or multiplex families in which two or more individuals are affected by schizophrenia. Three main assumptions are made. The first is that a major gene is segregating in at least some families and the second is that at least within families homogeneity can be assumed. In order to protect this second assumption some researchers have taken care to exclude families in which schizophrenia appears on both maternal and paternal sides. A focus on 'unilateral' pedigrees reduces the risk of having more than one major disease gene segregating within the pedigree. However, this selection process introduces a degree of bias into the third main assumption, which is that the mode of transmission in loaded pedigrees can be inferred at least approximately. Clearly if pedigrees of unilateral type with multiple members affected are selected, they will tend to have a dominant-like appearance and the assumption of a near-dominant mode of transmission has in fact been taken in practice by most recent The investigators of linkage in schizophrenia (Owen and McGuffin 1991).

The extent to which the assumptions in linkage studies of schizophrenia are safe have been discussed and explored by McGue and Gottesman (1991). They point out that on the basis of their simulation studies sampling multiplex families does not necessarily ensure that all individuals who are affected carry a major gene. If that gene happens to have a comparatively low frequency and a high heterozygote penetrance (i.e. the near-dominant model favoured by most investigators), then it is probable that not all affected individuals in multiplex families will carry the gene. However, if the gene frequency is high and the penetrance low, affected individuals in multiplex pedigrees will nearly always carry the major gene. Therefore it is almost certainly hazardous just to assume a near-dominant mode of transmission, and it is preferable to adopt a very broadly exploratory approach and apply a full range of models for the putative schizophrenia gene from completely dominant to completely recessive (McGuffin et al. 1983). Of particular interest is the range in models which would be compatible with published results on the distribution of schizophrenia in the population and in various categories of relatives of schizophrenics (O'Rourke et al. 1982). An even better strategy would be to study multiplex families forming the subset of a systematically ascertained series of families in which the sampling is based on consecutive probands presenting in a defined catchment area. The whole material could then be used in segregation analysis and the most plausible single gene parameters could subsequently inform the linkage analysis.

The whole issue of appropriate and efficient sampling strategies is of vital importance to the design of linkage studies. Goldin et al. (1991) have examined alternative sampling strategies ranging from a concentration on very large multiplex pedigrees to a focus on affected sibling pairs and their parents. In the middle are studies where nuclear families or pedigrees of moderate size are investigated. The affected sib-pair approach has attracted some investigators because of the simplicity of the analysis compared with likelihood approaches. In addition affected-pair methods are (nearly) 'model free' and therefore robust to the problems of misspecification of mode of transmission. However, sib-pair approaches lack power (Sturt and McGuffin 1985) and Goldin et al. concluded that they are generally less satisfactory than the analysis of families and large pedigrees. It was also of interest that the simulation studies of Goldin and colleagues show that linkage can be detected in the presence of heterogeneity in clinically realistic sample sizes. However, there are limits on this detection, so that, for example, heterogeneity presents a serious problem if the gene for schizophrenia is linked to the marker locus in less than 25% of families. Below this limit only very tight linkage can be detected.

The next practical decision to be made concerns what genetic markers to use and what regions of the genome to search. So far most have simply taken 'at random' whatever polymorphisms were available in the attempt to detect linkage. Indeed this was virtually all that could be done in the days when only classical markers were available. Nevertheless this 'random search' strategy has paid off with other disorders and again the example most relevant to psychiatry is Huntington's disease, in which linkage was chanced upon with one of eight random DNA polymorphisms (Gusella et al. 1983).

The prior odds against a marker selected at random being linked to a disease locus are high and a variety of methods can be used in an attempt to shorten the odds. The first and most obvious is to follow up on a previous positive result even if this presents only a hint of linkage. The second is to have some prior knowledge of the biochemistry of the disorder and to focus on loci which represent 'candidate genes'. In studies of schizophrenia this usually means genes encoding for proteins involved in neurotransmission, particularly those relating to dopaminergic systems. The third approach is to gain a clue from chromosomal anomalies and this has been the rationale for studies of chromosome 5 (Bassett et al. 1988) and, in part, for studies of the pseudoautosomal region of the sex chromosomes (Crow 1989). A combination of clues about potential candidate loci and observations on the cosegregation of mental disorders and chromosomal anomalies have led us to focus our own recent studies on a region on chromosome 11q. The preliminary results are interesting, and whether or not they ultimately stand the test of replication, they provide a useful illustration of an attempt to explore 'favoured region' of genome. Therefore we will describe briefly our chromosome 11q studies before moving on to discuss our findings in the context of previous and continuing linkage studies of schizophrenia.

Focusing on a Favoured Region: Chromosome 11q as an Example

The long arm of chromosome 11 is of potential interest for several reasons. First, several families have been reported in which there is apparent cosegregation of psychotic illness and balanced translocations involving this chromosome. The breakpoint in two of these is thought to be around 11q 21-q22 (Smith et al. 1989; St. Clair et al. 1990) while in the third the breakpoint is more distal at 11q 25 (Holland and Gosden 1990). Second, the human dopamine D_2 receptor (DRD2) has been mapped to chromosome 11q 21-q23. This is of interest because of the dopamine hypothesis of schizophrenia and in particular the suggestion from post-mortem ligand binding studies and in vivo positron emission tomography (PET) scan studies of increased densities of D_2 receptors in the brains of individuals with schizophrenia (Deakin 1988). Third, there has been a single report of a large family in which schizophrenic-like psychosis appeared to be coinherited with tyrosinase negative occulocutaneous albinism (Baron 1976), and it is now known that the tyrosinase (TYR) gene maps to 11q 14-q21. Lastly the gene for porphobilinogen deaminase (PBG-D) is located on 11q23-qTER (Wang et al. 1981). Acute intermittent porphyria results from mutations in this gene and may present with acute psychosis. Furthermore it has recently been suggested that one of the alleles of the Msp1 polymorphism at this locus is associated with schizophrenia (Sanders et al. 1991).

We therefore investigated 23 families multiply affected by schizophrenia, using a total of nine markers spanning our region of interest. Seven of the families came from South Wales, 13 from the southeast of England, 1 from southern Ireland and 2 from Japan. The family members were interviewed using either the PSE supplemented by the OPCRIT check list or the SADS-L and for the present analysis diagnoses were made according to the RDC. Two predetermined definitions of the phenotype were decided upon. The restricted definition included as affected all those suffering from schizophrenia, schizoaffective disorder and unspecified functional psychosis. The general definition was broader and included all those with the restricted phenotype plus family members receiving a diagnosis of schizotypal personality or major affective disorder (both unipolar and bipolar).

The markers were typed in our laboratories at the University of Wales College of Medicine and at the Institute of Psychiatry, London, and these are listed in Table 1.

As described above linkage analysis was performed assuming a range of models (McGuffin et al. 1983) for the putative schizophrenia locus from nearly complete dominant to recessive, including 'intermediate' models that are compatible with the distribution of schizophrenia in the general population and in published reports on various categories of relative (O'Rourke et al. 1982; Table 2). In addition the data were analysed using the affected family member method of Weeks and Lange (1988) and an 'affected only' lod score analysis was performed. All two point and multipoint analyses were carried out using the LINKAGE computer programs (Lathrop et al. 1985).

Table 1. Markers investigated on chromosome 11q

Locus	Marker type	Description	Typing method
INT2	AC repeat	MMT virus integration site	PCR
TYR	RFLP	Tyrosinase	SB
D11S35	AC repeat	Anonymous	PCR
DRD2	RFLP	Dopamine receptor D_2	PCR
CD3D	AC repeat	Antigen CD3D	PCR
PBG-D	2RFLP	Porphobilingon deaminase	PCR
D11S420	AC repeat	Anonymous	PCR

PCR, typed following amplification of genomic DNA by polymerase chain reaction;
SB, Southern blot analysis

Table 2. Some two allele single gene models of schizophrenia used in the analysis of 11q markers

Model	Penetrances			Frequency of A2
	A1A1	A1A2	A2A2	
Nearly dominant	0.001	0.95	0.99	0.02
Intermediate				
(i)	0.002	0.087	1.0	0.064
(ii)	0.0	0.061	0.656	0.04
Recessive	0.0	0.0	1.0	0.10

A2 is assumed to be the mutant ('disease') allele. Intermediate models assume a population risk of 1% and are compatible with published recurrence risks in relatives (McGuffin et al. 1983; O'Rourke et al. 1982)

The most interesing results concern the D11S35 locus, where a maximum 2-point lod score of 3.08 was obtained with the general clinical definition and an intermediate model of schizophrenia (model ii) which is compatible with the reported recurrence risks in the relatives of schizophrenics. By contrast both dominant and recessive models resulted in negative lod scores for this marker (Fig. 1). The intermediate model ii also produced positive but lower lod scores under the restricted definition of the phenotype. The only other positive results were with the Mbo I polymorphism at the TYR locus. The information here was low but weakly positive lod scores were found particularly with near-dominant models. The results of a multipoint analysis with schizophrenia as the test locus and TYR, D11S35 and DRD2 at fixed points are summarised in Fig. 2. Again these results were obtained using an intermediate genetic model and a general phenotypic definition. The maximum lod score is reduced to just below 3 and the most likely location of the schizophrenia locus is between TYR and D11S35.

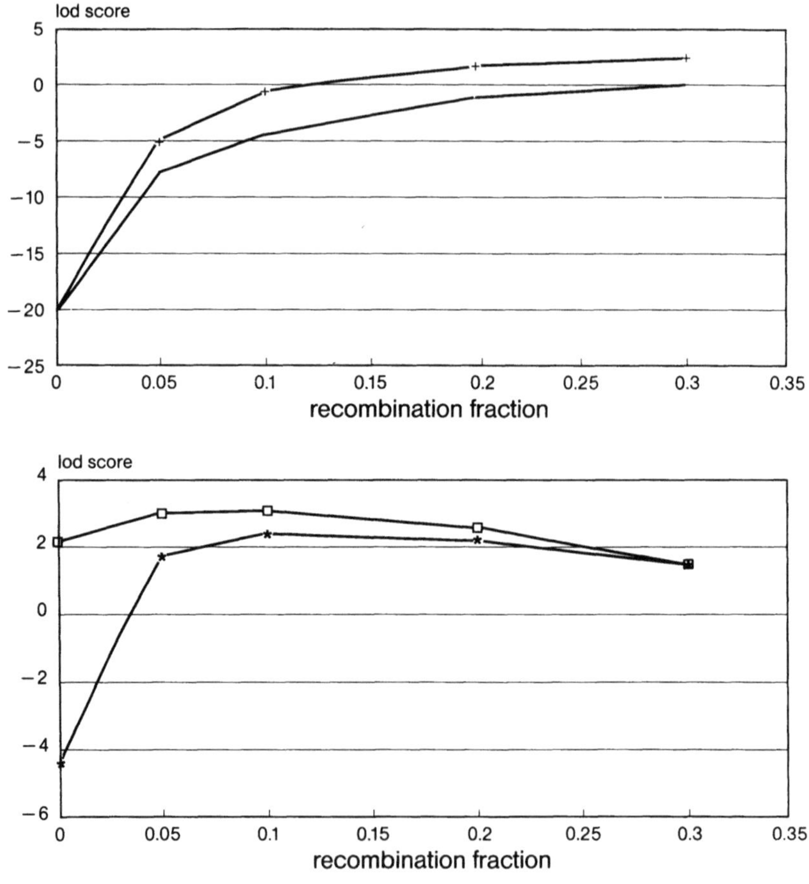

Fig 1. Two point lod score analysis of D11S35 and schizophrenia under (a) dominant —— and recessive —+—, and (b) intermediate models (intermediate I, —*— ; intermediate II, —□—)

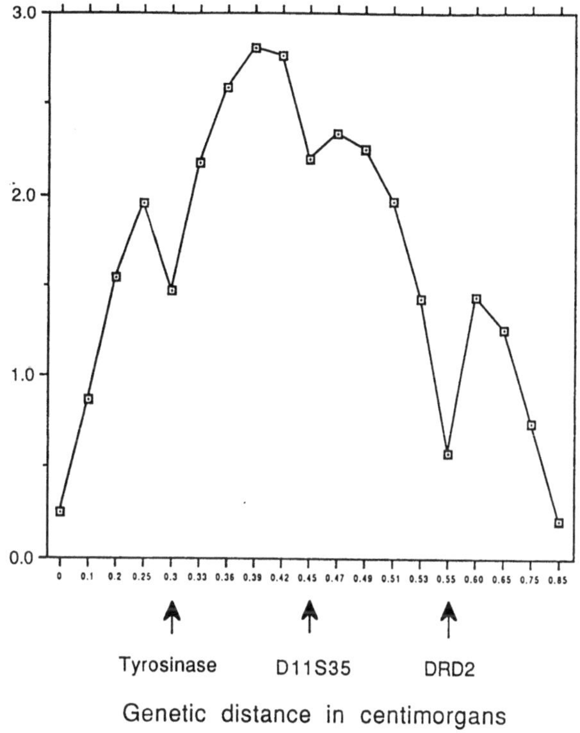

Fig. 2. Multipoint analysis, lod score, of three markers as fixed points on chromosome 11q with schizophrenia as the test locus

Heterogeneity analysis was performed on the basis of these multipoint data using the HOMOG program (Ott 1985) slightly modified by L. A. Sandkuyl (personal communication). HOMOG implements the admixture test of Smith (1963) and log likelihoods are computed for three competing hypotheses: H0, no linkage; H1, linkage with homogeneity; and H2, linkage with heterogeneity. The last of these had the highest maximum log likelihood and the likelihood ratio of H2 versus H1 was significant at only the 10% level (chi square 1.67; $p = 0.098$). Likelihood ratio tests also favoured H2 over the hypothesis of no linkage (chi square 12.92; $p = 0.0002$) and H1 over no linkage (chi square 14.59; $p = 0.003$). This can be interpreted as possibly suggestive evidence of linkage heterogeneity with an estimated alpha, or a proportion of linked cases of 0.75. Results with all other markers were uniformly negative and indeed linkage could be excluded for distances ranging from 10-25 centimorgans on each side of the highly informative microsatellite repeat markers INT2, D11S420, D11S490 and CD3D, depending on the genetic parameters specified in the analyses. Linkage analyses using haplotypes constructed from the two polymerase chain reaction (PCR) configured restriction fragment length polymorphisms (RLFPs) at the PBG-D locus also allow exclusion at genetic distances up to 20 centimorgans on either side for all genetic models.

Interpretation of the Findings

Our results suggest there is a gene which contributes to the liability to schizophrenia in a subset of families which is situated about 10 centimorgans centromeric of the marker D11S35 on chromosome 11q. Moreover, the highest lod score of 3.08 was found with analyses under a genetic model in keeping with the recurrence risks in various categories of relatives of schizophrenics; this model therefore appears to have more general plausibility and applicability than the ones from previous reports of positive linkage in studies in which it was assumed that schizophrenia segregates as a dominant or near-dominant trait in multiplex families (e.g. Turner 1979; Sherrington et al. 1988). However, our maximum lod score is only just over the level at which the evidence for linkage is conventionally accepted (Morton 1955).

A lod score of 3 corresponds to odds on linkage of 1000:1. However, because the prior probability of detecting linkage between two loci selected at random is low, of the order of 1:50, a lod score of 3 corresponds to a posterior probability of positive results arising by chance of 1 in 20, or a 'reliability' of 95% (Morton 1955; Edwards 1991). Testing multiple markers in a study of disease increases the prior probability that linkage will be detected, and therefore it can be argued that this is one instance where multiple testing actually reduces rather than increases the probability of type 1 errors (Ott 1985). Although at first sight paradoxical, this argument is probably correct (Risch 1991) so that the testing of multiple markers is unlikely to have given us a falsely inflated lod score. By contrast the effect of testing multiple genetic models and two different diagnostic classifications cannot be regarded as so benign. This exploratory procedure almost certainly will inflate the maximum lod score (Weeks et al. 1990; Clerget-Darpoux et al. 1990), and so it is prudent to apply a correction factor. A fairly conservative approach, suggested Risch (1991), is to subtract log 10 (t) where t is the total number of models tested. All told we explored eight genetic models and two diagnostic classifications so that our maximum lod score reduces by log 10 (16) to a modest 1.8. Nevertheless some support of our positive finding comes from the nonparametric analysis performed using the method of Weeks and Lange (1988). When allowance is made for low allele frequencies, the results are significant at the 0.02 level.

Clearly, at best, what we have discovered qualifies as a strong hint of linkage rather than proof that linkage exists between a schizophrenic gene and markers on chromosome 11q. Only an independent replication using the same methods would qualify as proof of linkage. Nevertheless it is interesting to ask what sort of gene this might be if indeed it does exist. It can be classed as a major gene in the sense that it is detectable using linkage methods. Although the term 'major' is usually left undefined, most people would probably interpret the term as signifying a gene which contributes all, or most, of the variance in liability to the disease. However, here we have a suggestion of an intermediate gene with low heterozygote penetrances, and when we calculate the total amount of variance accounted for using the usual formulae (James 1971), this amounts to only 21%. Thus there is plenty of room for the effects of other factors, including environmental influences, so that our results would be totally compatible with a multifactorial view of the aetiology of schizophrenia. Furthermore, the results of admixture analysis favour heterogeneity, with the gene on chromosome 11 being evident in about 45% of families. In short, we have a suggestion of a gene accounting for at least some of the variance and segregating in at least some of the families of schizophrenics. Where do these rather vague conclusions lead us, and how we should interpret them against a background of other studies to date?

Progress in Linkage Studies

Positive linkage findings in studies of schizophrenia are not a recent phenomena and indeed the first study using sib-pair analysis was published in the 1950s (Constantinidis 1958). Subsequently Elston and co-workers (1973) carried out a sib-pair analysis of dizygotic twins in which a variety of classical markers were typed and the results were suggestive of linkage with Gc, Rh and Gm. More recent studies on classical markers have effectively ruled out linkage with all three (McGuffin et al. 1983; Andrew et al. 1987). The first published attempt at detecting linkage in multiplex families and using a likelihood method of data analysis was a study of HLA types (Turner 1979). Here a very broadly defined clinical phenotype was taken which the author termed 'schizotaxia'. The analyses were performed under the assumption of a simple autosomal dominant mode of transmission resulting in a maximum lod score of 2.57 at a recombination fraction of 0.15. Four subsequent studies not only failed to replicate this finding but combining the results effectively enabled the exclusion of linkage between a dominant gene for schizophrenia and HLA up to a recombination fraction of 0.25 (McGuffin et al. 1983; Chadda et al. 1986; Goldin et al. 1987; Andrew et al. 1987). Negative results were also obtained when 'model free' affected sib-pair methods were employed to analyse the data. This suggests that misspecification of the mode of inheritance of the main trait did not account for the negative results. In addition further statistical analysis of the pooled data provided no support for the suggestion that there may be two forms of schizophrenia, one linked and the other unlinked to HLA (McGuffin 1988).

The most recent and most striking positive results were, as we have mentioned, in a study of British and Icelandic families in which a maximum lod score of 6.49 was obtained with a very broad phenotypic definition and a near-dominant model of schizophrenia in a multipoint analysis of markers in the 5q11-q13 region (Sherrington et al. 1988). Simultaneously a report of failure to find linkage in this area was published, based upon an investigation of a large Swedish kindred (Kennedy et al. 1988). However, since the Swedish family formed part of a relatively genetically isolated community, the possibility of linkage heterogeneity was raised (Kennedy et al. 1988; Lander 1988).

Subsequently several other studies have been published where attempts to replicate chromosome 5 linkage have proven negative. Studies of multiplex families from Scotland (St. Clair et al. 1989), Ireland (Diehl et al. 1989), the USA (Detera-Wadleigh et al. 1989; Aschauer et al. 1990) and Wales (McGuffin et al. 1990) have all failed to find linkage to markers in the same region of chromosome 5.

The discrepant findings resulted in considerable controversy since the original positive result appeared much more clear-cut and dramatic than any other previous single finding in this field. The possible reasons for the disparity between the one positive and all subsequent negative findings therefore require careful consideration. The first possible explanation is that the negative reports may be incorrect. There are, as we have mentioned, various difficulties associated with the linkage analysis of families with schizophrenia, resulting from such factors as diagnostic differences, incomplete penetrance, variable age at onset and genetic heterogeneity. These together with the risk of misspecifying the mode of transmission tend to militate against the finding of linkage even when this exists (Sturt and McGuffin 1985; Clerget-Darpoux et al. 1986). However, it is difficult to see why such factors would have operated uniformly in the negative studies and not in that of Sherrington and colleagues unless as a result of subtle differences in ascertainment and selection of families.

Secondly, nonallelic genetic heterogeneity might exist such that a defect on chromosome 5 accounts for only a minority of familial schizophrenias. Several human diseases have been shown to result from defects at more than one genetic locus and a recently studied example of psychiatric relevance is tuberous sclerosis (Sampson et al. 1989). The variation of symptom patterns in schizophrenia would support the hypothesis of aetiological heterogeneity but the cases studied by Sherrington and colleagues did not show obvious clinical differences from those studied in the negative reports, other than a much broader diagnostic definition in the most inclusive phenotype. Furthermore the fact that the sample studied by Sherrington and colleagues consisted of a mixture of British and Icelandic families makes it unlikely that differences in the gene pool explain the different results between their positive and the other negative studies.

Another possible explanation might be that chromosome 5 linked families were by chance overrepresented in the Sherrington et al. sample and underrepresented elsewhere. However, when the findings in all published studies in which sufficient information was available were reanalysed using a uniform set of genetic parameters (McGuffin et al. 1990), the results suggested that true heterogeneity was unlikely. An admixture test carried out on the multipoint scores from 34 families contributed by the five studies by Sherrington et al. (1988), Kennedy et al. (1988), St. Clair et al. (1989), Detera-Wadleigh et al. (1989), and McGuffin et al. (1990) did show significant evidence of heterogeneity. However, all the evidence came from the one positive British/Icelandic study, while the four other studies were uniformly negative for linkage with no evidence of heterogeneity. It therefore seems unlikely that true heterogenity exists since it would be expected that at least some suggestion of an admixture would have emerged when all of the negative studies were combined. With the exception of the Swedish pedigree of Kennedy et al. (1988) the other published reports consisted of several moderately large pedigrees rather than a small number of very large kindreds. Therefore one would expect a reasonable probability of at least a minority of these families containing the chromosome 5q linked form of schizophrenia, but there did not even appear to be a hint of such admixture.

Finally we need to consider the possibility that the findings of Sherrington and colleagues are incorrect either because of a systematic error or because they represent a chance positive finding. The most obvious source of systematic error would be if strict blindness had not been maintained between the investigators performing genetic marker analysis and those assigning psychiatric diagnoses. This is such a basic methodological point that it seems unlikely that it was overlooked. The second possibility is that division of samples could have occurred, leading to a false inference of linkage. Thus a large sample may show a suggestion of linkage in some families but not in others and it might be tempting to focus on the 'positive' families and set aside those that are not compatible with linkage. A series of moderately positive scores when summated would then give a misleading and spurious appearance of linkage. This again seems an unlikely explanation of the disparate findings.

We have already discussed the problems of multiple testing and it does seem likely that this may have falsely inflated the lod scores reported by Sherrington and colleagues. As reported in their published paper a total of four different definitions of the phenotype and six different genetic models were used in the linkage analysis. Therefore if we apply the correction factor suggested by Risch (1991) we would need to subtract log 10 (24) or 1.38 from the maximum lod score. This would still leave a score of 6.49 - 1.38 = 5.11, which would normally be regarded as strongly supportive of the hypothesis of linkage. Despite this, the failure of independent replication should probably be interpreted as excluding a schizophrenia gene from chromosome 5q for the reasons we have just discussed, and there is no satisfactory explanation of the single set of positive findings other than the possibility of a statistical fluke.

The only other published study with positive evidence of genetic linkage in schizophrenia relates to the pseudoautosomal region of the sex chromosomes. The hypothesis that this region might contain a gene for schizophrenia was based upon the observation that when two siblings are both affected by the disorder they tend more often than would be expected by chance to be of the same sex (Crow 1989). Further evidence for this proposal comes from the finding that same-sex concordance obtains only when the disease is paternally inherited (Crow et al. 1989). Molecular support for the hypothesis now comes from a study of sib pairs affected by RDC schizophrenia using a highly polymorphic genetic marker from the pseudoautosomal region (Collinge et al. 1991). This awaits replication but so far unpublished data from our own laboratories are negative. Given that the history to date of linkage marker studies of schizophrenia has consisted of a series of isolated positive findings followed by disappointment and refutation, how should we interpret our suggestive results involving chromosome 11q? The most immediate answer is 'with great circumspection'. However, there are several reasons for being reasonably sanguine. First, the positive lod scores were maximised with a model that is in keeping with published recurrence risks for schizophrenia in the relatives of index cases. This means that if the gene on 11q exists, it is of possible relevance to the transmission of schizophrenia generally and is not just a peculiarity of loaded pedigrees with a 'mendelian-like' appearance. Second, the putative gene on chromosome 11q accounts for only a modest proportion of variance.

The estimate of 21% quoted earlier cannot be viewed as dependable since, as we have discussed, our analyses were performed using a model of 'pure' single locus inheritance which is probably oversimplistic (McGuffin et al. 1985). Nevertheless the results are compatible with a multifactorial view of the aetiology of schizophrenia and are more plausible than invoking a near-dominant gene which accounts for nearly all of the variance. Third, the maximum support for linkage occurred when the definition of the phenotype was fairly broad although still restricted to severe mental illness. This is in keeping with the observation common in older studies (Slater and Cowie 1971) as well as some more recent investigations (Guze et al. 1983) that there is probably a slight excess of major depression as well as psychotic illness among the relatives of schizophrenics. Finally, there is a hint that linkage heterogeneity may be present within our sample and, again, this would be in keeping with a very long-standing intuitive proposition that genetic heterogeneity in schizophrenia is 'inherently probable' (Slater 1947; Gottesman and Shields 1982).

Other than attempted replication of the interesting preliminary findings in favoured regions such as that on 11q or the pseudoautosomal region, the way ahead in linkage studies must lie in systematic, coordinated, large-scale multicentre studies aiming to complete a thorough search of the genome. The practical and theoretical issues here have recently been discussed by Leboyer and McGuffin (1991). However, the success of such a time-consuming and expensive effort depends on the unproven hunch that genes of large effect exist. Since there is a finite possibility that they do not, it is worth considering other strategies.

Association and Alternatives to Linkage Strategies

Plomin and colleagues (1991) have recently recapitulated in stronger form some of the doubts of earlier authors that linkage studies have anything to offer in schizophrenia or other forms of abnormal behaviour. They suggest that such studies can only detect major gene effects "in which one gene is largely responsible for a behavioural disorder". This is probably overstating the case. Whether or not our own results on chromosome 11q prove to be correct, they demonstrate that it is theoretically possible for a linkage strategy to detect a gene accounting for only about a fifth of the variance in liability to schizophrenia. Nevertheless most would probably agree with Plomin and colleagues that the main purpose of a linkage study is in finding 'largely responsible' loci. As we have discussed, the search so far for genes of large effect involved in the transmission of severe mental illness is premised not on any clear evidence that such genes exist, but rather that there is no compelling evidence against their existence. By contrast there is compelling evidence that genetic effects of some sort are present in schizophrenia. Indeed such evidence constitutes the most important and consistent aetiological clue about the disorder (Gottesman 1991). The most comprehensive review of classical genetic studies has been provided by Gottesman and Shields (1982), who consider that a polygenic hypothesis offers the most satisfactory explanation of the transmission of schizophrenia. Plomin (1990) goes further and suggests that genetic influences on behaviours of any sort are much more likely to result from the combined action of multiple genes than from a single gene. He points to the evidence from human studies of cognitive and other traits and from animal breeding experiments in which heritabilities rarely exceed 50% and polygenic influences are favoured.

If these arguments are correct, then association strategies offer an approach to the molecular biological study of behaviour, which may ultimately be more fruitful than a programme devoted entirely to the search for linkage. Association studies may be capable of detecting minor susceptibility loci and are in most instances cheaper and simpler to perform than linkage studies. It is widely appreciated that linkage between loci does not normally result in allelic associations in populations because each generation contains a proportion of recombinants (r). Therefore after n generations the proportion of allelic pairs originally found in association has reduced to $(1-r)^n$. For example, if the recombination fraction between two loci is 0.1, then after twenty generations only 12% of alleles will remain in association. What is less widely appreciated is that where linkage is much tighter, association takes very much longer to break down (or return to equilibrium) so that a population study is more likely to discover so-called linkage disequilibrium. For example, at a recombination fraction of 0.01, after twenty generations 82% of the originally associated alleles will remain together. Another way of looking at this phenomenon is to consider the 'half-life' of an allelic association or the number of generations required to go half way to equilibrium (Morton 1982; Ott 1985). For unlinked genes this occurs after a single generation and at a recombination fraction of 0.1 it takes only 7 generations. However, the time taken is 69 generations for a recombination fraction of 0.01 and 692 generations when there is even tighter linkage and a recombination fraction of 0.001. The implication of these results is that once a very dense genetic linkage map has been developed, consisting of markers every 1 or 2 centimorgans, a systematic search for associations will become feasible.

What is the evidence that association strategies are capable of detecting genes of small affect? Probably the most convincing data are provided by studies of classical markers and disease. For example, even in apparently quite strong associations for which the relative risk of a disease for those with a certain HLA type versus those without is around 10 or more, HLA types only confer an increased susceptibility and never a certainty of developing the disease (Thompson 1981). Assuming a liability/threshold model of HLA-associated diseases in which liability is approximately normally distributed in the population, we have used the method proposed by Edwards (1965) to estimate the proportion of variance in liability to certain diseases conferred by HLA associations (Table 3).

Under the liability/threshold model even the best established and strongest association, that between HLA B27 and ankylosing spondylitis, accounts for only a small proportion of the variance. This is in keeping with the fact that the vast majority of B27-positive individuals in the general population never develop the disease.

In attempting to understand HLA associations more fully studies of the cosegregation of HLA and diseases such as insulin-dependent diabetes have been carried out in families. These produced initially perplexing results in which linkage did appear to be present but in which the recombination fraction between the disease and HLA

Table 3. HLA-disease association and proportion of explained variance in liability (EVL)

Disease/HLA type	Relative risk	EVL (%)
Ankylosing spondylitis/B27	82	12.9
Rheumatoid arthritis/DR4	4.1	6.3
Insulin-dependent diabetes/DR3	3.1	2.8
Multiple sclerosis/DW2	12	5.7
Haemochromatosis/A3	9.7	4.9

was too large to account for population associations (Hodge and Spence 1981; Clerget-Darpoux et al. 1986). The likely explanation is that what had been detected was not linkage in the traditional sense but a more complex phenomenon in which there is an epistatic interaction between HLA and another locus or loci. Interestingly this apparent 'linkage' is not confined to families segregating the HLA type found in association with diabetes mellitus in the population. That is, HLA seems to represent a marker for a susceptibility locus to insulin-dependent mellitus which can only account for a small proportion of the variance in liability in the population. Nevertheless the effect is large enough to perturb independent assortment in families. Suarez and colleagues (1982) have pointed to a similar phenomenon in multiple sclerosis. They have calculated that the HLA relationship only accounts for about 3% of the variance but still produces a consistently detectable distortion of the identity by descent scores for HLA types of affected siblings.

It therefore seems quite possible that studies of multiplex families in other complex diseases, including schizophrenia, will detect loci which are not strictly speaking major and from which the results cannot be strictly interpreted as demonstrating 'linkage'. If the analogy with HLA and disease findings is carried further a relationship with polymorphic DNA markers should be demonstrable in population samples. As we have mentioned, this will of course require a human genetic linkage map which is much more complete than that currently available. For example, a map consisting of 1500 evenly spaced markers would mean that any disease susceptibility locus would have a recombination fraction with at least one marker locus of about 0.01 or less, and there might then be a reasonable prospect of detecting allelic association resulting from linkage disequilibrium. The same broad strategy might be applied to continuously distributed characteristics, and a preliminary search for quantitative trait loci for cognitive traits has already begun (Plomin et al. 1991). We can conclude therefore that the arguments for performing association studies are persuasive, particularly in the long term. However, any positive results which do emerge will not lead on as neatly to reverse genetic strategies as they have done already for some single gene disorders.

References

American Psychiatric Association (1980) Diagnostic and statistical manual of mental disorders (3rd edn) (DSM- III). APA, Washington DC

Andrew B, Watt DC, Gillespie C, Chapel H (1987) A studyof genetic linkage in schizophrenia. Psychol Med 17:363-370

Aschauer HN, Aschauer-Treiber G, Isenberg RD et al. (1990) No evidence for linkage between chromosome 5 markers and schizophrenia. Hum Hered 40:109-115

Baron M (1976) Albinism and schizophreniform psychosis: a pedigree study. Am J Psychiatry 133:1070-1073

Bassett A, McGillivray BC, Jones BD, Pantzar JT (1988) Partial trisomy chromosome 5 cosegregating with schizophrenia. Lancet 1:799-801

Bloom FE (1990) Strategies for understanding the role of gene defects in the pathogenesis of mental disorders. In: Bulyshenkov V, Christen Y, Prilipko L (eds) Genetic approaches in the prevention of mental disorders. Springer, Berlin Heidelberg New York, pp 45-56

Chadda R, Kullhara P, Singh T, Sehgal S (1986) HLA antigens in schizophrenia. A family study. Br J Psychiatry 149:612-615

Clerget-Darpoux F, Bonaiti-Pellie C, Hochez J (1986) Effects of misspecifying genetic parameters in lod score analysis. Biometrics 42:393-399

Clerget-Darpoux F, Babron M-C, Bonaiti-Pellie C (1990) Assessing the effect of multiple linkage tests in complex diseases. Genet Epidemiol 7:245-253

Cloninger CR, Reich T, Yokoyama S (1983) Genetic diversity, genome organisation and investigation of the etiology of psychiatric diseases. Psychiatr Dev 3:225-246

Collinge J, Dehisi LE, Boceio E et al. (1991) Evidence for a pseudo-autosomal locus for schizophrenia using the methods of affected sibling pairs. Br J Psychiatry 158:624-629

Constantinidis JK (1958) Les marquers de chromosomes chez les schizophrenes et la recherche du linkage entre ces caracteres et la schizophrenie par la methode de Penrose. J Genet Hum 7:189-242

Crow TJ (1989) A pseudo-autosomal locus for the cerebral dominance gene. Lancet 2:339-340

Crow TJ, Delisi LE, Johnstone EC (1989) Concordance by sex in sibling pairs with schizophrenia is paternally inherited: evidence for a pseudoautosomal locus. Br J Psychiatry 155:92-97

Davies K, Read AP (1988) Molecular basis of inherited disease. IRL, Oxford

Deakin JFW (1988) The neurochemistry of schizophrenia in schizophrenia: In: Bebbington P, McGuffin P (eds) The major issues. Heinemann, Oxford, pp 56-72

Detera-Wadleigh SD, Goldin LR, Sherrington R, Encio I, de Miguel C, Berrettini W, Gurling H, Gershon ES (1989) Exclusion of linkage to 5q11-13 in families with schizophrenia and other psychiatric disorders. Nature 340:391-392

Diehl S, Su Y, Aman M, Machean C, Walsh D, O'Hare A, McGuire M, Kidd K, Kendler K (1989) Linkage studies of schizophrenia in Irish pedigrees. Paper presented at the First World Congress on Psychiatric Genetics, August 1989, Cambridge

Edwards JH (1965) The meaning of the associations between blood groups and disease. Ann Hum Genet 29:77-83

Edwards JH (1991) The formal problems of linkage. In: McGuffin P, Murray R (eds) The new genetics of mental illness. Butterworth Heinemann, Oxford, pp 58-70

Egeland JA, Gerhardt DS, Pauls D et al. (1987) Bipolaraffective disorders linked to DNA markers on chromosome 11. Nature 325:393-399

Elston RC, Kringlen E, Namboodri KK (1973) Possible linkage relationship between certain blood groups and schizophrenia or other psychoses. Behav Genet 3:101-106

Farmer AE, McGuffin P, Gottesman II (1987) Twin concordance for DSM-III schizophrenia: scrutinising the validity of the definition. Arch Gen Psychiatry 44:634-641

Farmer AE, McGuffin P, Harvey I, Williams M (1991) Schizophrenia; how far can we go in defining the phenotype? In: McGuffin P, Murray R (eds) The new genetics of mental illness. Butterworth Heinemann, Oxford, pp 71-84

Feighner JP, Robins E, Guze SB, Woodruff RA, Winokur G, Munoz R (1972) Diagnostic criteria for use in psychiatric research. Arch Gen Psychiatry 26:57-63

Goldin LR, De Lisi LF, Gershon ES (1987) The relationship of HLA to schizophrenia in 10 nuclear families. Psychiatry Res 20:69-78

Goldin L, Martinez MM, Gershon ES (1991) Sampling strategies for linkage studies. Eur Arch Psychiatry Clin Neurosci 240:182-187

Gottesman II (1987) The psychotic hinterlands or the fringes of lunacy. Br Med Bull 43:556

Gottesman II (1991) Schizophrenic genesis. W H Freeman, San Francisco

Gottesman II, Shields J (1982) Schizophrenia: the epigenetic puzzle. Cambridge University Press, Cambridge

Gusella JF, Wexler NS, Conneally PM et al. (1983) A polymorphic DNA marker genetically linked to Huntington's disease. Nature 306:234-8

Guze SB, Cloninger R, Martin RL, Clayton PJ (1983) A follow-up and family study of schizophrenia. Arch Gen Psychiatry 40:1273-1276

Harrison PJ, Pearson RCA (1989) Gene expression and mental disease. Psychol Med 19:813-820

Hodge SE, Spence MA (1981) Some epistatic two-locus models of disease II. The confounding of linkage and association. Am J Hum Genet 33:396-406

Holland A, Gosden C (1990) A balanced chromosomal translocation partially co-segregating with psychotic illness in a family. Psychiatr Res 32:1-8

James J (1971) Frequency of relatives for an all-or-none trait. Ann Hum Genet 95:47-59

Kendler KS, Gruenberg AM, Strauss JS (1981) An independent analysis of the Copenhagen sample of the Danish adoption study of schizophrenia. II. The relationship between schizotypal personality disorder and schizophrenia. Arch Gen Psychiatry 38:982-984

Kennedy JL, Giuffra LA, Moises H, Cavalli-Sforza LL, Pakstis AJ, Kidd JR, Castiglione CM et al. (1988) Evidence against linkage of schizophrenia to markers on chromosome 5 in a northern Swedish pedigree. Nature 366:167-170

Kety SS, Rosenthal D, Wender PH, Schulinger F, Jacobsen B (1976) Mental illness in the biological and adoptive families of individuals who have become schizophrenic. Behav Genet 6:219-225

Lander ES (1988) Splitting schizophrenia. Nature 366:105-106

Lathrop GM, Lalouel JM (1984) Easy calculations of lod scores and genetic risks on small computers. Am J Hum Genet 36:460-465

Lathrop GM, Lalouel JM, Julier C et al. (1985) Multilocus linkage analysis in humans: detection of linkage and estimation of recombination. Am J Hum Genet 37:482-498

Leboyer M, McGuffin P (1991) Collaborative strategies in the molecular genetics of the major psychoses. Br J Psychiatry 158:605-610

McGuffin P (1988) Genetic markers: an overview and future perspectives. In: Smeraldi E, Bellini L (eds) A genetic perspective for schizophrenia and related disorders. Edi-Ermes, Milan, pp 145-156

McGuffin P, Festenstein H, Murray RM (1983) A family study of HLA antigens and other genetic markers in schizophrenia. Psychol Med 13:31-43

McGuffin P, Farmer AE, Gottesman II, Murray RM, Reveley AM (1984) Twin concordances for operational definitions of schizophrenia. Arch Gen Psychiatry 41:541-545

McGuffin P, Farmer AE, Gottesman II, Murray RM, Reveley AM, (1985) The appropriate use of criteria for schizophrenia. Arch Gen Psychiatry 42:423-424

McGuffin P, Sargeant M, Hett G, Tidmarsh S, Whatley S, Marchbanks RM (1990) Exclusion of a schizophrenia susceptibility gene from the chromosome 5q11-q13 region. New data and a re-analysis of previous reports. Am J Hum Genet 47:524-535

McGuffin P, Farmer AE, Harvey I (1991) A polydiagnostic application of operational criteria in studies of psychotic illness: development and reliability of the OPCRIT system. Arch Gen Psychiatry 48:764-770

McGue M, Gottesman II (1991) The genetic epidemiology of schizophrenia and the design of linkage studies. Eur Arch Psychiatry Clin Neurosci 240:174-181

McGue M, Gottesman II, Rao DC (1985) Resolving genetic models for the transmission of schizophrenia. Genet Epidemiol 2:99-110

Morton NE (1955) Sequential tests for the detection of linkage. Am J Hum Genet 7:277-318

Morton NE (1982) Outline of genetic epidemiology. Karger, Basel

O'Rourke DH, Gottesman II, Suarez BK et al. (1982) Refutation of the single locus model in the aetiology of schizophrenia. Am J Hum Genet 33:630-649

Ott J (1985) Analysis of human genetic linkage. Johns Hopkins University Press, Baltimore

Owen M, McGuffin P (1991) DNA and classical genetic markers in schizophrenia. Eur Arch Psychiatry Clin Neurol Sci 240:197-203

Plomin R (1990) The role of inheritance in behavior. Science 248:183-188

Plomin R, McClearn GE, Owen M, McGuffin P (1991) Quantitative genetics, molecular genetics and dimensions of normal variation (submitted)

Risch N (1991) A note on multiple testing procedures in linkage analysis. Am J Hum Genet 48:1058-2064

St. Clair D, Blackwood D, Muir W, Baillie D, Hubbard A, Wright A, Evans HJ (1989) No linkage of chromosome 5q11-q13 markers to schizophrenia in Scottish families. Nature 339:305-309

St. Clair D, Blackwood D, Muir W, Carothers A, Walker M, Spowart G, Gosden C, Evans HJ (1990) Association within a family of a balanced autosomal translocation with major mental illness. Lancet 336:13-16

Sampson JR, Yates JRW, Pirrit LA, Fleury P (1989) Evidence for genetic heterogeneity in tuberous sclerosis. J Med Genet 26:511-516

Sanders AR, Hamilton JD, Chukraborty R, Fann WE, Patel PI (1991) Association of genetic variation at the porphobilinogen deaminase gene with schizophrenia. Am J Hum Genet (submitted)

Sherrington R, Brynjolfsson J, Petursson H et al. (1988) Localization of a susceptibility locus on chromosome 5. Nature 336:164-167

Slater E (1947) Genetic causes of schizophrenia sysmptoms. Monatsschr Psychiatr Neurol 113:50-58

Slater G, Cowie V (1971) The genetics of mental disorders. Oxford University Press, Oxford

Smith LAB (1963) Testing for heterogeneity of recombination fractions in human genetics. Ann Hum Genet 27:175-182

Smith M, Wasmuth J, McPherson JD, Wagner C, Grandy D, Civelli O, Potkin S, Litt M (1989) Cosegregation of an 11q22 3-9 p22 translocation with affective disorder: proximity of the dopamine D2 receptor gene relative to the translocation breakpoint. Am J Hum Genet 45:A220

Spitzer RL, Endicott J, Robins E (1978) Research diagnostic criteria for a selected group of functional disorders, 3rd edn. New York State Psychiatric Institute, New York

Sturt E, McGuffin P (1985) Can linkage and marker association resolve the genetic aetiology of psychiatric disorders: review and argument. Psychol Med 15:455-462

Suarez B, O'Rourke D, Van Eerdwegh (1982) Power of the affected sub-pair method to detect disease susceptibility loci of small effect: an application to multiple sclerosis. Am J Med Genet 12:309-326

Thompson G (1981) A review of the theoretical aspects of HLA and disease associations. Theor Popul Biol 20:168-208

Turner WJ (1979) Genetics markers for schizophrenia. Biol Psychiatry 14:177-205

Wang AL, Arredondo-Vega FX, Giampetro PF, Smith M, French Anderson W, Desnick RJ (1981) Regional gene assignment of human porphobilinogen deaminase and esterase A4 to chromosome 11q23-11qter. Proc Natl Acad Sci USA 78:5734-5738

Weeks DE, Lange K (1988) The affected-pedigree member method of linkage analysis. Am J Hum Genet 42:315-326

Weeks DE, Lehrner T, Squires-Wheeler E, Kaufman C, Ott J (1990) Measuring the inflation of the lod score due to its maximization over model parameter values in human linkage analysis. Genet Epidemiol 7: 237-244

Wing JK, Babor T, Brugha T, Burke J, Cooper JE, Giel R, Jablenski A, Regier D, Sartorius N (1990) SCAN. Schedules for clinical assessment in neuropsychiatry. Arch Gen Psychiatry 47:589-593

Discussion of Presentation P. McGuffin et al.

Blackwood
Let me just comment on the D11S35 -linkage. Replication is the best defense against possible causes of false positives. Unfortunately we do have nonreplication in 17 schizophrenic families. We get lod scores of about -5, -6, between 5% and 10% recombination. So it is not replicated. On the other hand, I think it is an interesting possibility for addressing heterogeneity. I have one question. The difference you get in shifting models is really puzzling: you get a lod score of -5 with a purely recessive model and dominant mode, but when you go to an intermediate model, which is not all that different, you suddenly get +3, which makes a lod score difference of 8. Is that huge shift likely on simulation studies?

McGuffin
Yes, it is likely on simulation studies. Indeed in recent work of Clerget-Darpoux it was shown that you can get big shifts in the lod scores in actual complex diseases, depending on which model you apply, going from negative to positive values. I take the other point, and of course we know about your negative data, because the first thing we did when we found these results was to ask who in the UK would have similar data. Thus we asked you to analyze your data in a similar way and you kindly did. What we should do now is actually to pool the data together and see whether heterogeneity could be the cause of the apparently differing results.

Vogel
You gave a fairly optimistic view on the chances of association studies, especially if the loci are relatively closely together. However, you also have to remember that such association studies may give you absolutely negative results if you assume that multiple mutations may have produced this special genetic variant of the marker you have been using and that there is a large, random mating population. In this case you may find completely negative results with the association studies despite the fact that there might be linkage with the gene, which contributes a relatively high proportion of the variation you are looking for. Therefore, I would be a little bit more cautious in this respect.

McGuffin
Yes, there are the difficulties you have mentioned, but on the other hand, there are also the difficulties with association studies of false positives, which for me are actually much worse difficulties, because the prior probability of finding association is very low. If we want to correct in association studies for the low prior probability like one does in linkage analysis one might come up with quite different *p* values.

Ott
I think one should use a p value of 0.001. You know a lod score of 3 has the p value at most being equal to 0.001. It will be comparable.

Vogel
Just a remark: You may have two different reasons for an association. The one reason might be a linkage disequilibrium, being relatively close to linked loci, between which there is no physiological relationship. The other possibility is that the locus you are studying itself is influencing the phenotype in question. When you look at John Edwards' paper on the association of blood group O and duodenal ulcer in this case the assumption is that the ABO locus has actually an influence on the genetic variation leading to ulcer. There is a physiological relationship. In this case, of course, all your remarks are correct. However, if it is only due to linkage disequilibrium with the two loci, with no physiological relationship at all except that they are just closely linked by chance, then you may get these negative results very often.

McGuffin
Yes, I think the only mathematical assumption in John Edwards' paper is that an association shifts the liability curve over to the right and that may be because of linkage disequilibrium or because of the effect of the locus itself. But, I agree that certainly most of the speculations on the reasons for blood group-disease associations have concerned what I call pleiotropic effect.

Mendlewicz
You mentioned that you used a not too narrow and not too broad definition of the phenotype. Could you tell us a little bit more about that?

McGuffin
We used two diagnostic definitions. The first one was RDC schizophrenia plus schizoaffective disorder plus unspecified psychosis of a nonaffective type. We call that restricted definition. The second and third of those categories, the schizoaffective and unspecified psychoses, did not account for very many subjects. The general definition, as we refer to it in the written paper, also included schizotypal personality disorder and affective psychoses. Now, there is not actually an affective psychosis category in RDC, only unipolar depression. But in fact all of the unipolar relatives had a psychotic depression.

Barnard
On a similar topic, you mentioned that your analysis excludes an association of schizophrenia and chromosome 5. But can you comment on the translocation evidence of the cases of Basset.

McGuffin
I think Basset's observation was interesting and all of the negative chromosome 5q studies do not rule out the fact there may still be something etiologically important there. We have not ruled out an association on chromosome 5q; we ruled out a linkage and we think after having reanalyzed the data published worldwide, including Gurling's data, that we were able to rule out the controversial 5q linkage.

Tsuang
I would like to ask about the boundaries of the definition of schizophrenia. You mentioned that you used the schizoaffectives, both depressed and manic. And also in the general definition you include the affective psychoses. So could you elaborate a little bit more. Does this mean that you think schizophrenia and affective psychosis genetically are of a similar genotype?

McGuffin
In general, no. But within families, where there are several schizophrenics and someone has an affective psychosis, my opinion is that the person with affective psychosis does not have anything different from the others.

Tsuang
So the result which you are reporting on 11q is actively based on very broad definition.

McGuffin
It is not very broad. The lifetime risk of all of those disorders together in the general population is not going to be more than 2%-3%.

Tsuang
After this discussion we may need to talk about the boundaries of schizophrenia and affective disorders, too. Particularly, we are more interested in unipolars among the relatives of bipolar probands. And as you know, many unipolars have nothing to do with the affective disorders. And some unipolars may be different from bipolars, and only a fraction of unipolars may really have the genotye of bipolars. So in general, I think for the discussion we may need to come to the definition of the phenotype at the different levels for genetic research.

McGuffin
I agree with you. I think the trouble with some of the modern concepts of unipolar disorders is that you just have to have a hint of depression and you have got unipolar

disorder. These are severely ill people in these families and I think they all have got the same illness.

Reich

It seems most likely that the wider definition casts the phenotype beyond a core phenotype the more likely genetic heterogeneity, both within and between families, is to occur. That is true with schizophrenic psychoses as affective illnesses. With that in mind, you do have more cases and, in fact, do you have more genetic cases. It seems to me likely that one would lose rather than gain power by widening the phenotype.

McGuffin

I agree with that also, but I would like to see whether we can get the same effect that we got when with Irv Gottesman, Anne Farmer, and I looked at the twin concordance for schizophrenia. We found that when we added extra categories over and above DSM-III schizophrenia, we got an increase in the MZ to DZ concordance ratio, but it reached a peak with the addition of atypical psychosis, schizotypical personality and affective disorder with mood incongruent psychosis. When we went on adding other categories, including unipolar disorder and any Axis-I diagnosis, the MD to DZ ratio plummeted. What I would like to do is find a linkage that can be replicated and then optimize my diagnostic definition. We would then hope to see a peak in the evidence for linkage at the optimal diagnostic definition and show a decrease in evidence with further broadening of the definition.

Kidd

Two comments, one about Ann Basset's data. It's very hard to evaluate how significant that finding is. It came about because it brought to attention a family with two affected. If you take away the prior unlikelihood of two schizophrenics in a familiy, and one of them might be dysmorphic, then really the fact that the two relatives shared schizophrenia and this translocation has a prior probability of only about one fourth by chance alone, which means it is not terribly strong evidence for a clear etiologic relationship. And with respect to the last comment, quite a few years ago when one of my students was first analyzing the importance of including age at onset effects, we analyzed the linkage data in some of the early reports for color blindness and affective illness. Given that there was a set of data already known to be positive, we could not replicate it. It was interesting that as we broadened the phenotype, the lod scores went up when we went from bipolar and added unipolar, but when we added alcoholics, the lod scores plummeted. So it is a very real question of where one draws the boundary in these diagnoses and I certainly don't know the answer. I am not convinced anybody knows the answer, and it is clearly a question we have to address.

Maier

I would like to ask you about your graph and the variance explained in dependency of the penetrance of the heterozygotes. In the model specified here, you have a rather low penetrance. In homozygotes, A2A2, there is only 0.66. I wonder if this would not also have a substantial impact on the reduction of the variance. I would expect from this figure that this is the major reason for this.

McGuffin

No, it is not. If you simply take the formula from James' 1971 paper in *Annals of Human Genetics*, and you take the penetrance vector and value of your gene frequency and you can plot graphs like this fairly easily. I plotted a very simple one with F1 and F3 penetrance fixed and varying the F2. The surprising result is that when you get into those really intermediate areas with very low heterozygote penetrance you are not explaining much variance.

Reich

I was intrigued about the idea that, if there were various forms of illness under evolutionary pressure, the forms that would survive would have to account for the least variance in that graph. If the penetrances were higher, disease genes should be under more selective pressure. There is no logical reason to expect that penetrances have to be below 20% in the heterozygotes. However, that kind of mutant gene would be able to persist in the population at fairly high frequency, despite selection pressure.

Gurling

We have studied more families on chromosome 5 in Iceland and in England and these are negative. So it seems likely that our original result was a false positive. But just to be certain, we looked at the original cohort of families, and they still have positive lod scores above 3 overall. We thought there might be a possibility that this family, which shows a sort of dominant transmission with incomplete penetrance, might indicate involvement of chromosome 5. We have localized the 5-$HT1_A$ receptor gene very close to one of our markers and found an RFLP with this neurotransmitter receptor. However, our results with this RFLP appear to be false positives, too.

Clinical Observations Critical for Linkage Studies of Schizophrenia and Affective Psychoses

C. R. Cloninger

Introduction

The recent failures of replication in linkage trials of psychiatric disorders such as schizophrenia and affective psychoses are likely to be a consequence of the fact that the mode of inheritance of these disorders is unknown. The design and analysis of linkage trials has been inadequate for psychiatric disorders that are as common and developmentally complex as schizophrenia and affective psychoses. In this paper I will summarize some clinical observations that are critical in the design and analysis of linkage trials for complex psychiatric disorders and then offer some specific recommendations to improve the replicability of future studies.

Mode of Inheritance

Family, twin, and adoption studies clearly demonstrate the importance of genetic factors in schizophrenia and affective psychoses and provide important information for the design of linkage studies. Information is even more complex for schizophrenia than for affective psychoses, so I will focus on schizophrenia in considering what we can conclude about its mode of inheritance. The relevant data have already been summarized in detail elsewhere (Cloninger 1988).

First, only about 20% of schizophrenic individuals have a first-degree relative with broadly defined schizophrenia. Accordingly, multiplex families are not representative of schizophrenia in general. Care must be taken in identifying multiplex schizophrenia pedigrees to exclude families of atypical affective psychoses, especially families of schizoaffective bipolar probands, which are more likely to have densely loaded pedigrees. In the recently designed National Institute of Mental Health molecular genetic initiative and our ongoing linkage studies of schizophrenia (Aschauer et al. 1990), we require that the proband satisfy DSM-III-R criteria for schizophrenia and that at least one other nuclear family member have schizophrenia or schizoaffective disorder with a drepressive syndrome (i. e., schizoaffectives with a manic syndrome are acceptable as probands).

Second, most of the cotwins of a schizophrenia proband do not have broadly defined schizophrenia and over 30% are clinically normal with neither psychosis nor personality disorder. This discordance can be explained by incomplete penetrance or sporadic phenocopies. Studies of the risk of schizophrenia in the adult offspring of monozygotic twins discordant for schizophrenia reveal about the same risk of schizophrenia in the children of the discordant cotwin as in the children of the schizophrenic twin. Likewise the risk of schizophrenia in first-degree relatives of monozygotic pairs discordant for schizophrenia is the same as in relatives of concordant twin pairs. This shows that the penetrance of schizophrenigenic genotypes is incomplete, but phenocopies are infrequent once coarse brain disease has been excluded. Extension of pedigrees must allow for incomplete penetrance by extending through unaffected parents who have parents or sibs who are themselves affected.

Third, schizophrenigenic genotypes have a highly variable expression. The risk of schizophrenia and schizoaffective disorder is the same in the first-degree relatives of probands with either DSM-III-R schizophrenia or DSM-IIII-R schizoaffective disorder. Accordingly, the DSM-III-R cirteria for schizophrenia are too narrow or restrictive for use in genetic analysis of pedigrees. Furthermore, in the first-degree relatives of subjects with other nonaffective psychoses (such as delusional disorder and atypical nonaffective psychoses) there is a substantial incidence of narrowly defined schizophrenia. Accordingly, other nonaffective psychoses represent a second level in the schizophrenia spectrum in terms of the certainty or probability that they are expressions of an underlying schizophrenigenic genotype. In addition, schizotypal personality and affective psychoses with mood-incongruent delusions are also part of the "schizophrenia spectrum" as judged by their familial association with narrow schizophrenia. Improved ways of systematically and quantitatively accounting for variable expressivity will be discussed in the next section.

Fourth, quantitative genetic analyses of schizophrenia pedigree data have only been able to exclude the highly restricted model of a single major locus that causes all cases of schizophrenia and has no familial environmental effects. This leaves open almost any mode of inheritance, including a single major locus with special environmental effects for twins, locus heterogeneity (i.e., different major genes, any one of which is sufficient to cause schizophrenia), oligogenic inheritance (i.e., a few genes which each contribute substantially to susceptibility but are usually not sufficient to cause schizophrenia alone), and multifactorial inheritance (i.e., a large number of genes and/or enviromental factors which each contribute slightly to the risk of schizophrenia). Most linkage trials have recently been conducted as if it were most likely that there were one or more major genes sufficient to cause schizophrenia, that is, under the assumption of single major locus inheritance or locus heterogeneity. However, single locus inheritance, locus heterogeneity, or additive polygenic inheritance cannot explain the high concordance of schizophrenia in monozygotic twins compared to the concordance in first-degree relatives. The high concordance in monozygotic twins requires allowance for either nonadditive gene-gene interaction (i.e., epistasis) or environmental effects unique for monozygotic twins. The rapid reduction in risk for schizophrenia from identical twins to second-degree and third-degree relatives is most suggestive of epistasis in the absence of degenerate ad hoc assumptions about special environments for each degree of genetic relationship. In other words, the most likely and parsimonious model of inheritance for schizophrenia is an oligogenic model with epistasis, not locus heterogeneity. Such a model of inheritance has some strong implications for improving the design and replicability of linkage trials of psychoses.

Semicontinuous Quantification of Clinical Spectra

Prior linkage studies have handled variable clinical expression by testing a series of overlapping definitions, usually increasingly broad and inclusive. This trial-and-error approach has two serious limitations. First, it increases the likelihood of false-positive results because of multiple comparisons if uncorrected; or, correction for multiple comparisons reduces significance levels and power to detect linkage. Second, multiple categorical definitions fail to utilize available quantitative information about liability to disease. Family studies (ideally adoption studies of separated biological relatives) provide a direct quantitative assay of the liability to a disorder represented by any particular subgrouping of a disease spectrum. This is a problem for which the multiple threshold model of disease transmission was developed by Reich and his colleagues (Reich et al. 1975). Let us consider the diagnoses of schizophrenia, schizotypal personality, psychotic depression with mood-incongruent delusions, and alcoholism. Suppose the morbidity risk in first-degree relatives for schizophrenia in the relatives of each group of probands is 6% schizophrenia in the relatives of schizophrenic probands, 4% schizophrenia in the relatives of schizotypal probands, 2% schizophrenia in the relatives of mood-incongruent depressive psychosis probands, and 0.5% schizophrenia in the relatives of alcoholics and in the general population. The relative risk of schizophrenia for each type of proband is 12 for schizophrenia, 8 for schizotypal personality, 4 for mood-incongruent depressive psychosis, and 1 for alcoholism. This provides a quantitiative scale for assessing the probability or certainty that an individual has a schizophrenigenic genotype. Such semicontinuous rankings can be derived using relatively model-free assumptions by logistic regression or under more specific models of inheritance. Basically, each diagnostic class can be rated in terms of extent of deviation from the population mean for its ability to transmit or to be transmitted by a disorder.

Furthermore, such ratings for syndromes that occur in the same families can be refined by information about individual symptoms, severity of incapacitation, age at onset, numbers of hospitalizations or episodes, or biological markers. This has already been shown useful to characterize stability of diagnosis by Rice et al. (1987). All that is required to incorporate such information is family data about the relationship of the putative variables. This is likely to reduce to some extent changes in diagnosis with continued follow-up of families if subthreshold clinical features and biological markers are taken into account in rating the probability that an individual has a putative genetic susceptibility.

In addition, information about the sensitivity and specificity of available information can also be taken into account. Individuals who are young and not through the period of risk, or about whom medical records or direct interviews are not available, cannot be rated with the same certainty as older individuals with more complete information. Accordingly, we should really rate the probability and certainty of an individual having a putative genetic susceptibility rather than try to artificially categorize an individual as affected or unaffected.

Ideally, quantitative information about such semicontinuous ratings should be derived for the sample under study by a particular set of raters in a particular setting. That is, the ascertainment of the sample should be sufficiently explicit and systematic that segregation analysis can be carried out along with the linkage analysis. However, this would require more systematic ascertainment of families from well-defined sampling frames than has been characteristic of linkage studies in the past. Other reasons for more systematic ascertainment of pedigrees for linkage studies of psychoses will be considered in the next section.

Ascertainment of Pedigrees for Oligogenic Disorders

In many recent linkage trials pedigrees have been ascertained as if the mode of inheritance were known or as if there were only a single major locus or locus heterogeneity at worst. Nonsystematic opportunistic selection of pedigrees was carried out to find large pedigrees with a high density of illness. Such selection should permit detection of linkage, but does not permit correction for ascertainment for segregation analysis. If there is only a single locus contributing to risk in a particular pedigree, misspecification of the gene frequency and penetrance functions for the susceptibility locus does not prevent the detection of linkage, but may substantially bias the estimate of the recombination fraction in two-point analyses between the marker locus and the putative susceptibility locus (Ott 1985). Furthermore, addition of flanking markers for multipoint analyses may only lead to false exclusion of the entire region when the parameters of a susceptibility locus that is truly in the region are misspecified. If there are two or more loci contributing to risk in a pedigree, proper specification of the penetrances of the susceptibility genotypes cannot be readily approximated by a putative single susceptibility locus, particularly if there is epistasis and the contributions for each locus are roughly equal. Accordingly, in order to map susceptibility loci for oligogenic disorders, it is necessary to know the mode of inheritance (including genotype frequencies and penetrances of the multilocus genotypes) or to be able to do combined segregation and linkage analysis. Otherwise false exclusion of the entire genome may occur by multipoint analyses in which the relevant susceptibility genotypes are misspecified. Basically the mapping of susceptibility loci for oligogenic disorders, especially in the presence of epistatic interactions among multiple loci, requires systematic ascertainment and extension of pedigrees.

Furthermore, simulation studies of oligogenic disorders (i.e., those involving 2 to 6 loci) by Brian Suarez and his colleagues (personal communication, 1991) have found that when linkage for a major locus is detected, attempts at replication will often fail. Some ascertainment rules are worse than others in terms of replicability. In particular, the common practice of rigidly excluding pedigrees with any suggestion of bilineality produces pedigrees that appear to be autosomal dominant with affected individuals in successive generations. Selection for such structure leads to highly unreplicable linkage claims. Accordingly, replicability is likely to be enhanced by allowing inclusion of affected relatives on both sides of a pedigree, if these occur naturally.

This suggests that linkage trials of psychoses are likely to proceed in two stages. In the first phase, which we are now undergoing, nonsystematic ascertainment of pedigrees will lead to reports of the detection of linkage and mapping susceptibility loci if the disorders are mendelian or involve only single locus heterogeneity; this phase will lead to reports of detection of linkage that are often unreplicable if the disorders are oligogenic. In the second phase, systematic ascertainment of pedigrees for both segregation and linkage analysis will be needed to map susceptibility loci for oligogenic disorders.

If resources are available, systematic ascertainment would be wise because of the likelihood that the psychoses are oligogenic disorders. Another consideration is the high-cost and long-term effort involved in doing even nonsystematic ascertainment. Extensive genome searches and lifelong follow-up of pedigree members mean that there is no such thing as a cheap and quick linkage study of psychiatric disorders, given the delayed and variable age at onset in these individuals. Accordingly, we should conduct linkage studies as systematically as possible in order to deal with the complexities of possible multigenic inheritance.

Conclusions

The mode of inheritance of psychotic disorders such as schizophrenia and affective psychosis is characterized by incomplete penetrance, delayed and variable age at onset, and variable expression. The most likely mode of inheritance for schizophrenic disorders, for which the most information is available, is more likely to be oligogenic inheritance with epistasis than single major locus inheritance or locus heterogeneity. In contrast, most recent linkage studies have been designed as if only single genes contributed substantially to illness in individual pedigrees. This assumption has led to approaches to ascertainment which are likely to lead to unreplicable results for oligogenic disorders. Improved replicability would follow from systematic ascertainment and extension of pedigrees. This would facilitate combined segregation and linkage analysis, which is critical to unbiased estimation of recombination fractions in linkage analysis and therefore crucial for mapping of susceptibility loci in oligogenic disorders. Reduction of false claims of linkage and false claims of exclusion of linkage will also be facilitated by quantitative estimation of the probability that an individual has a psychotigenic genotype, taking into account all available signs, symptoms, course information, and information on source of information. Available techniques make systematic ascertainment more rigorous and ultimately more cost-effective than nonsystematic approaches to mapping susceptibility genes for complex disorders.

References

Aschauer HN, Aschauer-Treiber G, Isenberg RD, Cloninger CR (1990) No evidence for linkage between chromosome 5 markers and schizophrenia. Hum Hered 40:109-115

Cloninger CR (1988) Genetics of schizophrenia. In: Kaplan HI, Sadock BJ (eds) Comprehensive textbook of psychiatry, 5th edn. Williams und Wilkins, New York, pp 732-744

Ott J (1985) Analysis of human genetic linkage. John Hopkins University Press, Baltimore

Reich T, Cloninger CR, Guze SB (1975) The multifactorial model of desease transmission. Br J Psychiatry 127:1-32

Rice JP, Endicott J, Knesevich MA, Rochberg N (1987) The estimation of diagnostic sensitivity using stability data: an application to major depressive disorder. J Psychiatr Res 21:337-345

Discussion of Presentation C. R. Cloninger

Ott

You assume that we have a linkage and that this linkage has been established in different cases. But people sometimes try to choose different models and look to see which one gives the highest lod score. How would you see this procedure as opposed to the other one, where you select families in a specific fixed way?

Cloninger

Clearly I think we should stop trying to maximize lod scores. That is too tantalizing. But, if we shift the focus more to experiment-wide estimates of significance level, correcting for multiple comparisons, that will tend to inhibit people simply arbitrarily searching on gene frequencies and penetrances in order to maximize the lod score. I personally am very impressed by the strategy for detecting linkage that Neil Risch has advocated and which we saw applied by Dr. McGuffin: that is, to consider a recessive and a dominant, and in McGuffin's case, an intermediate case, trying all three possibilities as a screening step for detecting linkage. This seems to have been a rather robust and sensitive approach in the Genetic Analysis Workshop. At the current stage, until we have very large, systematically ascertained pedigrees, in which we are able to do both segregation and linkage analysis, I think that would be what I recommend. I am sure there are other opinions.

Maier

You recommended using probability statements instead of arbitrary dichotomies for defining phenotypes and probably all agree to this proposal, even given the fact that nearly no linkage study proceeded in that way. But there might be a lot of disagreement with regard to the basis of this probability statement. You cited the paper by John Rice. He used diagnostic stability as a main criterion. But when discussing schizophrenia, it is very clear that the syndrome or the symptom which has highest diagnostic stability are the schneiderian first rank symptoms, and we know from twin studies that they have no heritability.

Cloninger

I would not use stability as a measure of familiality. I would use the influence on morbid risk in relatives of items, or biological markers, or syndromes. I think the general principle of using logistic regression to estimate weights, using concordance in relatives instead of concordance in follow-up is a very analogous procedure.

Maier

What do you think of using the heritability estimate from twin studies as probability assessments?

Cloninger

It would be comparable. That would be very closely related. The question is just what scale should be used. What mode of inheritance should be assumed to estimate heritabilities? I would just stick to analysis as model free as possible and use something like logistic regression.

Reich

I just wanted to let people know briefly that Ming Tsuang, Bob Cloninger, and myself are part of a consortium of American investigators involved in a National Gene Bank Program. These investigators have agreed to develop an instrument which is specifically designed with many of the points in mind that Bob Cloninger and Miron Baron have made to carry us through to a new way of genetic linkage studies in psychiatry. The instrument is almost at the final stage. It is called DIGS (Diagnostic Instrument for Genetic Studies); there is a companion family history instrument called FIGS. It is polydiagnostic, it can fit into the OPCRIT program, and it does phenotyping of individuals in terms of severity. It further attempts to make spectrum diagnoses. We intend to do follow-ups so we may find out something about stability of diagnoses as well. A test of within and between center test-retest reliability was done after training interviewers in a group at seven centers; preliminary data showed that it behaved very well. We should be confident of the diagnoses made within or between centers. So at least the first properties of the instrument are in place. The intention of the program is to generate approximately 200 families of schizophrenic probands with approximately 1600 relatives and cell cultures and 250 families of bipolar probands with a few more relatives and cell cultures as well. The ultimate intention of gathering this very large data set is to provide a data base for the scientific community, including DNA and various diagnoses. I think the common bond of the group is that all of the groups in Europe, Canada, and the US were looking for the same genes and attempting to replicate each other's results. Furthermore, we believe that if we have very precise control, as precise control of all experimental variables as possible, then the chances of getting replicable results are amplified, whatever the mode of transmission and whatever the model is. It suggests some kind of standardization procedure.

Cloninger

The approach that we are taking in the US does differ from the one you are taking in the European Science Foundation (ESF) collaboration. I wonder if some of those investigators could comment on how they feel about the recommendations I made earlier about systematic ascertainment and extension of pedigrees.

Mendlewicz

Of course we have witnessed a lot of instruments in psychiatry and psychopathology in the last 20 years. Thus I wanted to ask you, why one more instrument? What is the specific target, what are the specific objectives that would make this instrument a more reliable one than the ones already existing?

Reich

The properties that we sought were as follows: Firstly, that the diagnostic criteria of the major psychoses are ambiguous. For example, an individual diagnosed with schizoaffective manic disorder by RDC or modified RDC criteria can have schizophrenia but not affective disorder by DSM-III-R diagnosis. The same individual and the same family, for example, might be used in both studies, based on the alternately used criteria. It seems mandatory, especially with ICD-10 coming along with DSM-IV, to have an instrument that is sufficiently broad and covers enough of the bases, so information can accumulate over the years rather than have to be dropped. Further, that to the extent measurements are made using various criteria sets, they can be used to provide population data and other relevant parameters as well. No single instrument fulfilled even that basic criterion.

In addition, we thought about the phenomenon of the atypical or unusual case. Most simple diagnostic instruments do not enable or encourage the detection of atypical cases, so that they can be excluded or labeled and treated separately during a linkage analysis, hopefully to eliminate or reduce the possibility of false positives. Finally, probably over half of the patients with affective or schizophrenic psychoses, who are clinical cases, have comorbid anxiety or substance abuse disorders. I am especially pointing to the strong affinity of cocaine abuse and bipolar manic depression. With that in mind, we developed a separate section, which still specifically deals with issues of comorbidity in order to better understand and delineate that phenomenon in linkage analyses. These are some of the properties we have built into it, which perhaps represent a new generation of measurements that we feel are relevant.

Mendlewicz

I think it is a very ambitious project. I agree that the available instruments are not doing the best job. We know that they have many problems and the question is why should geneticists be more successful than the expert biometricians.

Reich

My private belief is that the clinical geneticist, the one who actually does family studies, has a different task to perform than, for example, the epidemiologist.

Gurling

How could systematic ascertainment help overcome the heterogeneity problem, the locus heterogeneity problem? It seems to me that probably the biggest problem in schizophrenia will be heterogeneity, and we don't know how many subtypes in schizophrenia there are going to be. But if it is like diabetes; there could be 10 or quite a number of subtypes. Systematic ascertainment of large pedigrees might best resolve the issue of heterogeneity.

Cloninger
Not neccessarily. We should try to get a range of a large and moderate-sized pedigrees, so that we do have an opportunity to do some analyses within an individual pedigree in the event there is locus heterogeneity. However, we don't know whether we are dealing with several genes that are each sufficient to cause the disorder or with several genes which interact nonadditively with each other to influence the susceptibility. The problem is that no ascertainment procedure is optimal for all different modes of inheritance. So, to characterize the mode of inheritance, you are going to have to be able to do both segregation and linkage analysis. You are not going be able to do that in large pedigrees that are not systematically ascertained.

Gurling
I would like to defend the approach that the ESF has adopted. Because we will get the very best families in Europe, we can then study those cases we are absolutely certain about and do not have to include phenotypes which are less certain.

Cloninger
I think that is a fine strategy as long as you have several different genes that each cause the disorder in question. But if you have anything more complicated than that, such as epistasis, you are going to be "up a creek without a paddle" after you get past the stage of detecting. You may not be able to carry it further.

Mendlewicz
We do have indeed a project in Europe which is being supprted by the European Science Foundation. For those who may not be familiar with the ESF, perhaps Peter McGuffin could say a few words about the ESF and the European approach.

McGuffin
I agree in many ways with all that Bob Cloninger was saying about having to have a systematic ascertainment as well as having the big pedigrees and concentrate on linkage analyses. But, in the ESF collaboration, which takes in about 16 countries and where the funding is available for the core aspect, the workshops and the communication, we have to take into account our own national and international heterogeneity in Europe. We also have to take into account money. A systematic ascertainment approach is very expensive. Similarly, I have sympathy with all Ted

Reich's criticism of the currently available diagnostic interviews. But, developing a new interview is expensive and time-consuming and that is why in Europe we have taken a very pragmatic approach and taken PSE-10 and the lifetime SADS and adopted them as the acceptable interviews. But, we have the OPCRIT program, which you can define your data into, too, and this will enable you to have both a polydiagnostic approach and comparably cheap interviews. It would be nice to have a more luxurious, well-funded approach. Using OPCRIT data we can combine the European and US data sets.

Cloninger

We would also agree that these different strategies are complementary. What you are doing may well pay off more quickly, at least in the detection of linkage. And I think that association studies may also generate some results, but we are going to hear from Dr. Kidd about the pitfalls of that. So everything has its strength and its weakness, but I am concerned about nonsystematic ascertainment.

Mendlewicz

I may just add that there is a collaborative task force. The American and the European groups are meeting to try and find a common basis for the exchange of information and data. So I hope that we can have a joint project, where at least we have a common basis for exchanging and communicating the data.

Reich

I would like to distinguish the systematic aspects of ascertainment in pedigree extension from simply having rules that one follows. It may well be that the rules that one follows may change in time. However, it seems to me that it is very important to have rules, even if you have 16 sets of them. But at least the investigator should know the rules that they play by, so that other people can replicate the results as precisely as possible. I would also encourage making available DNA, audiotapes, and other phenotypic measurements, so an independent confirmation of the results can be taken and audited as well.

Tsuang

I think the definition of spectrum disorders among the family members is very crucial at this time for the studies of the genetics of schizophrenia. There is no argument with regard to definition of probands, how to select the typical probands, but once you have selected the proband, how do you get into the definition of spectrum disorders. It seems to be very important to draw the boundaries. As you know, DSM-III schizotypal personality disorder, as an example, is actually a mixture of everything, a combination of the positive symptoms and some negative symptoms. In fact, those patients who are seen by the clinician as schizotypal are those who come to visit the clinic, and they don't have a familial basis. And those with the negative symptoms seldom come to see clinicians, but they are more familial. So, recently we looked into which components are more important. It is actually a flatness of affect and the avolition. These two, even in the absence of the positive symptoms, seem to play a very important role. I mentioned this because Cloninger suggested quantitative grading and probability. Currently we rate the relatives of schizophrenia patients, comparing them with the relatives of patients with affective disorders, and the negative symptoms of the rating are significantly different from those of the relatives of depression, which are mostly negative symptoms flatness of affect and avolition. Positive symptoms actually are very unstable and do not have a familial basis. The reason I mention this is that we probably should not take schizophrenia as the one disease category to study genetics. We may have to look into each subitem of the symptom manifestations and look into those symptoms which are more in character with the basis of a core familial basis. So the time may have come when genetics can contribute to the development of the nosology of schizophrenia and vice versa. And this is very important if we want to make a breakthrough in this area.

The Complexities of Linkage Studies for Neuropsychiatric Disorders

K. K. Kidd

Within the last decade there has been considerable fluctuation in opinion of whether genetic linkage was an appropriate paradigm for identifying major genes of psychiatric disorders and, even if appropriate, how such studies should be designed. The reviews presented by Baron (this volume) and by McGuffin et al. (this volume) are quite comprehensive and cover aspects of the history of the field as well as current perspectives on the major scientific issues and difficulties facing researchers in this field. While there are some who view the possibility of finding genes for psychiatric disorders by genetic linkage to be highly problematic at best, others, myself included, view it as the single most promising avenue of approach to these complex disorders. A promising line of research, however, is not necessarily simple and linkage studies of complex neuropsychiatric disorders are among the most difficult. Nevertheless, the effort is worthwhile given the individual suffering and extraordinary public health problems presented by these disorders. The variety of problems, difficulties, unresolved issues, and uncertainties raised by Baron (this volume) and McGuffin et al. (this volume) need not be restated. Rather, in the following paragraphs I will present a slightly different perspective on some of the relevant issues, provide some different examples relating to points they made, and, in at least one case, present a slightly different opinion.

The Value of Knowing the Genetic Map

Human genetics prior to 1978 can be called the dark ages of modern human genetics. In 1978, the first DNA polymorphism, identified directly in the DNA molecules, was discovered by Kan and Dozy (1978). This was a restriction fragment length polymorphism (RFLP), a term which refers to the particular technology for visualizing these differences (Botstein et al. 1980). There are now additional technologies for detecting DNA polymorphisms which cannot be called RFLPs, but the term is often used to refer to all DNA polymorphisms. Until 1978, there were no known DNA polymorphisms; by 1981 there were 23 loci with known DNA polymorphisms. By the Tenth Human Gene Mapping Workshop in 1989 almost 2000 DNA marker loci had been cataloged (Kidd et al. 1989). The number of mapped and cataloged DNA markers has now increased to above 2400 (Kidd 1991). During the last decade, the total has increased 100-fold, from two dozen to more than 2400 of these normal genetic variants (markers) in *Homo sapiens* that are identifiable in the DNA. Today human geneticists can use these to follow through families virtually any part of any chromosome to observe exactly how it has been transmitted; we can easily answer questions such as which grandchildren inherit which genes from which of the grandparents. It is this extraordinary new ability to know how relatives share genes that makes linkage studies not only feasible but powerful.

Until the last 2 or 3 years, there were very few accurate maps of these markers. Maps are rapidly being assembled now but much of the genome still lacks accurate genetic maps even though there are now markers in almost all regions: finding markers and roughly mapping them is far different from obtaining an accurate genetic map with both correct order along the chromosome and reasonably accurate distances among the markers. The accurate genetic maps that are being assembled using this plethora of genetic variation will be an important resource for linkage studies of complex neuropsychiatric disorders. Knowing the genetic map will allow us to test for linkage using simultaneously a set of markers spanning a defined region of a single chromosome rather than testing markers and analyzing them one at a time. With a simultaneous search approach (multipoint analysis, Lathrop et al. 1984) much more linkage information is extracted than with a traditional two-point analysis because it increases the chances of a meiotic product being informative for the segment of the chromosome being studied. Complete specification of how the DNA has actually been transmitted through a family is an important aspect of linkage studies for complex disorders. However, the results of multipoint analysis are extremely dependent upon the accuracy of the map being used in conjunction with the analysis. For example, one of our (Heutink et al. 1991) severe criticisms of the analyses published by Brett et al. (1990) was that the logarithm of odds (lod) score was obtained using a map with the markers in the wrong order.

Knowing the genetic map also allows one to build up a gradual exclusion map for a locus so that future efforts can be targeted most efficaciously at the regions not yet studied or excluded because of data in only one family. Recent exclusion maps for major disorders are those for Tourette syndrome with over 80% of the genome excluded (Pakstis et al. 1991a and unpublished observations) and for affective disorders in the Old Order Amish with over 25% of the genome excluded (Pakstis et al. 1991b). However, these exclusion maps are quite sensitive to misspecification of the mode of inheritance (the genetic model) for the trait (Risch and Giuffra 1991). Thus, what is actually excluded is not just any major gene but a major gene of the specific nature hypothesized and used in the analyses.

Candidate Genes in Their Genomic Context

The importance of candidate genes seems overstated with respect to the complex neuropsychiatric disorders. Our knowledge of genes relevant to the structure and function of the central nervous system is increasing at an extraordinarily rapid pace but still is abysmally small. Estimates from a decade ago (Sutcliffe 1988) still appear valid: between one third to one half of all genes function primarily to exclusively in the central nervous system. One would have to be very lucky if one of the few such genes currently identified were to be the etiologically relevant gene. Nonetheless, there is no reason to avoid studying such genes and it can be done profitably in the context of scanning the genome using the genetic map and multipoint linkage analysis. Such an approach allows one to extract maximum information on segregation of the chromosomal segment containing the "candidate" gene by simultaneously utilizing other markers on either side.

Our study of the dopamine D2 receptor locus (DRD2) is a case in point. Though now it is much more informative for linkage with multiple polymorphisms distributed along the gene (Sarkar et al. 1991; Hauge et al. 1991), initially there was only a single TaqI polymorphism of moderate heterozygosity. We used that polymorphism and other markers nearby to assemble a genetic linkage map for the region around DRD2 and then used those markers and that map in studies of Tourette syndrome (Gelernter et al. 1990) and schizophrenia (Moises et al. 1991). We obtained a much more definitive exclusion of DRD2 as the etiologically relevant locus in both of those studies as a result of the nearby markers. Moreover, we have now excluded a segment of the long arm of chromosome 11, a small but noticeable increment to the exclusion maps for these loci.

A cautionary note to the candidate gene approach using linkage is illustrated by the recent discovery that the gene for the amyloid precursor protein is the cause of Alzheimer's disease in several families (Goate et al. 1991). A linkage study had previously excluded this gene based on obligate crossovers in some families. Genetic heterogeneity may be the explanation for the crossover data but, whatever the explanation, direct sequencing of the gene from patients finally identified the defect in the amyloid precursor protein in a family that had not shown a crossover with that locus. While it does not explain Alzheimer's disease in all families (see Van Broeckhoven, this volume), this mutation does represent a significant advance in understanding Alzheimer's disease.

Positional Cloning

Our focus in the current situation is on identifying by linkage a region of the genome of etiologic significance for each disorder. However, that is not the ultimate goal. Eventually we want to identify the actual gene involved and understand its normal and abnormal function. Linkage is simply the first step in the positional cloning strategy that has recently been so successful for genetic diseases such as cystic fibrosis (Riordan et al. 1989; Rommens et al. 1989; Kerem et al. 1989) and the fragile X mental retardation (Yu et al. 1991; Oberlé et al. 1991). Thus, our objective is not to exclude 100% of the genome, but to find a region containing a relevant gene.

Linkage Analysis, Lod Scores, and Significance Tests

In family material the traditional method for a linkage analysis has involved utilization of maximum likelihood with significance determined by the lod score, basically the logarithm of a likelihood ratio. Based on the prior likelihood of two loci chosen at random being on completely separate chromosomes, Morton (1955) initially chose a lod score of +3 as the desired significance level. However, among pairs of loci giving a lod score of +3, 1 in 20 (5%) will be false positives. From any laboratory where extensive linkage analysis for DNA polymorphisms is being done, mine included, several examples of lod scores greater than three for loci that are not located on the same chromosome can be cited. If this were the only consideration, it is not too surprising that many researchers would want a higher level of certainty when an important disease is at issue. The question of multiple tests becomes a very significant one during intial stages of a linkage study because each locus tested is essentially unlinked to all other markers. Thus, by chance alone, at least one of those is more likely to surpass a lod score of +3 than if only a single locus were being tested. A simple correction is to increase the lod score for significance by $\log_{10} n$, where n is the number of independent markers tested (Kidd and Ott 1984). A lod score of +5.3 would therefore suffice for the first 200 loci being studied on a random basis. As more markers are studied or if markers are studied in linkage groups, the situation is more complex because the segregation patterns of some of the markers are correlated whenever those markers are linked to each other. The Tourette Syndrome Association Consortium on Genetic Linkage has adopted a lod score of +5 as its minimal value. Certainly, in the search for linkage for Tourette syndrome we have identified several individual markers giving lod scores with Tourette syndrome that were greater than +3. Those positive lod scores mostly disappeared when more information was collected in the same families using additional closely linked markers.

However, correcting lod scores for multiple tests, whether multiple markers or multiple diagnostic schemes, is not one of the primary concerns when interpreting data on a complex disorder. One must also recognize that the analysis is parametric and so based on requisite *assumptions* (Kruger et al. 1982). The primary concern is that the statistical premise for use of lod scores in this situation is wrong — the phenotype is not stable.

A linkage analysis compares two hypotheses: whether the phenotypes in the pedigree can be better explained (a) by assuming linkage when inferring patterns of segregation of alleles at the hypothesized trait locus and at the marker loci or (b) by assuming independent segregation. Note, one infers transmission patterns for both a hypothetical trait locus and the marker loci. One step can be taken to shore up this virtual house of cards: making the transmission patterns for the marker loci as unambiguous as possible by studying multiple highly informative markers; many such are available and more are becoming available. That step does not address the real issue, however, since the transmission pattern inferred for the trait locus can change dramatically with a single change of phenotoype. Consequently, not only do we not know what to use as a significance level, but it is also unclear whether statistical significance in the conventional sense is relevant.

The lod score does not address the issue of what the evidence would be when the next person in the family becomes affected. Most of the disorders being considered have variable and occasionally quite late onset. While this can be built into the likelihood equations being used for the linkage (Morton and Kidd 1980), those equations do not deal with the issue of the next onset. On the hypothesis of linkage, the age-at-onset distribution and the segregation of the linked marker will predict which individuals are most likely to develop the disorder. However, on the null hypothesis that the trait and marker are not linked, any individual in the pedigree might be affected and only the age of the individual may be a relevant factor in prediction. If, as happened in the Old Order Amish between the time of the original positive lod score (Egeland et al. 1987) and the reevaluation a few years later (Kelsoe et al. 1989), the "wrong" person becomes affected, the inferred segregation pattern for the hypothesized trait locus may be drastically altered and the lod scores affected far more than one might have otherwise expected.

Rather than rely on lod scores alone, of whatever significance level, one might consider the alternative of relying on the stability of the lod score to changes in diagnosis. One method for evaluating the stability of a lod score in a given family is a series of reanalyses of the data with, in turn, each of the unaffected members being reclassified as affected. This would generate a distribution of lod scores and could be used to determine how confident one is in a given value. We have seen individual families for which existing data gave a lod score of slightly greater than +2 while a single new onset generates a distribution ranging from a lod score of only slightly greater than +2 (the initial lod score) to values considerably below zero. Thus, moderately positive evidence can change into negative evidence with only one additional onset. The converse of this problem, the difficulty raised by a false positive — a phenocopy or etiologically different form of the disorder being considered as affected within the pedigree — can be addressed by the converse "simulation": generating a distribution of lod scores if each affected individual were, in turn, considered to be of unknown phenotype. This approach can help identify those individuals in a data set who are most critical for any given conclusion. For example, if a negative lod score suddenly becomes moderately positive when one affected individual is "rediagnosed" as unaffected or unknown, attention should be paid to the diagnostic characterization of that individual. Such non-blind, hypothesis-driven attention to diagnosis of single individuals is not without its own risks. So long as one is aware that these are post hoc evaluations for the sake of generating new, testable hypotheses, it is perfectly acceptable. Indeed, the ultimate validation of any major gene involvement in a neuropsychiatric disorder will not come from linkage analysis but from a detailed molecular and physiological understanding of the gene and its function. The linkage analyses only serve to help us find the genes; they will not be the definitive studies.

Typing Markers Blind to Diagnosis

It seems unnecessary to worry about blindness of laboratory personnel in studies of clear-cut linkage markers. In the context of a genome survey and multipoint analyses the marker typings are validated by marker to marker linkage studies and ultimately by the unambiguous identification in all segments of the pedigree of virtually all crossover events that have been preserved in the genotypes of descendants.

Association Studies

Association studies have long been used in attempts to understand the etiology and pathogenesis of complex neuropsychiatric disorders such as alcoholism. Such studies compare the frequency of some characteristic or biochemical variant in patients with the frequency of that same trait or measurement in a sample of unaffected individuals. If the measured characteristic is involved in the etiology or pathogenesis of the disorder, the frequency or mean value should be different in the two samples.

When an association study is done with a DNA polymorphism, a slightly, but significantly, different hypthesis is being tested than when a biochemical marker is being tested. Since most DNA polymorphisms represent normal variation in DNA sequences that have no known function, a direct causal relationship between the alleles studied and the disorder is not likely. Instead, the studies are directed at finding relevant functional variation indirectly through its being somehow associated with the DNA polymorphism. Such an association study is therefore dependent on linkage disequilibrium: the relevant functional variation is distributed nonrandomly (in the population of chromosomes) with regard to the observed DNA polymorphism.

There is fundamentally only one reason for such an association, though the situation may have arisen in several ways. The simplest would be that in the past a single mutation at the functionally relevant site occurred on a specific chromosome with a particular DNA sequence at the polymorphic site. The two sites are sufficiently close that recombination does not occur between them except very rarely. Subsequently, chromosomes with the functionally relevant mutation and, hence, that paricular allele at the DNA polymorphism are the only ones giving abnormal susceptibility to the disease. Alternatively, a variety of population genetic processes may have resulted in chromosomes with one particular allele at the polymorphic site, representing a large fraction of the chromosomes with abnormal susceptibility to the disease.

In contrast, there are many reasons why an association might not be found even if the functionally related variation is close to the DNA polymorphism: (1) Multiple mutations to the disease susceptibility state would occur on a random set of chromosomes with a random sampling of the sequences at the DNA polymorphic site. If there have been enough different mutations to high susceptibility, the frequencies of alleles at the DNA polymorphism will be the same on high susceptibility chromosomes as on low susceptibility chromosomes. Many disease loci exhibit this pattern. (2) Recombination has occurred frequently enough between the two sites that, even if there is only one mutation, alleles for the DNA polymorphism occur on "abnormal" chromosomes in roughly the frequencies they occur on "normal" chromosomes. The polymorphisms at the 5' end of the β-hemoglobin cluster show this "scrambling" with respect to the few independent mutations to the sickle cell hemoglobin (Antonarakis et al. 1984). (3) Only a few mutations have occurred, but their frequencies and the particular polymorphism alleles with which each is associated are such that when the disease-producing mutations are not distinguished and instead pooled, the frequencies of polymorphic alleles within that set of chromosomes are essentially the same as among normal chromosomes. The phenylalanine hydroxylase polymorphisms on normal and phenylketonuria chromosomes show this last pattern in the Danish population (Kidd 1987; Daiger et al. 1989).

Much has been made of the power of association studies with genetic markers but the strongest, clearest examples are of associations with HLA antigens. Such associations are special cases because the antigens themselves are important functional components of the immune system and because there is known to be strong linkage disequilibrium across many hundreds of kilobases of DNA in the HLA region. Thus, even if an association does not directly reflect etiologic involvement of the HLA antigens, it may indirectly reflect the very high amounts of linkage disequilibrium in this region. For there to be an association of a DNA polymorphism in the region of a "candidate gene" and a complex neuropsychiatric disorder, one must assume that linkage disequilibrium is the explanation since virtually all DNA polymorphisms are demonstrably in nonfunctioning segments of DNA — the non-coding intervening segments and flanking regions of genes. (Of course, some of those regions might have unknown regulatory functions as yet unidentified, but for a variety of reasons that seems unlikely for most polymorphisms.) Linkage disequilibrium is not uniformly distributed along the DNA molecules since recombination events are not uniformly distributed. Thus, one must determine empirically whether or not such disequilibrium could exist given the location of a polymorphism with relationship to the functional components of a gene. Even if disequilibrium could exist, it is a priori extremely unlikely for a common disorder and statistical tests need to consider this prior likelihood.

Conclusion

Despite their problems when used to study complex disorders, linkage studies remain the best hope for identifying major genes causing these disorders. However, because the possibility of false-positive findings is larger than has been generally assumed, greater care must be taken in future work not to overinterpret results of linkage studies. Even stronger cautionary statements can be made about association studies which are unlikely to yield useful results unless there is exceptionally strong prior evidence (e. g., biochemical data) that a specific gene causes a disorder.

References

Antonarakis SE, Boehm CD, Serjeant GR, Theisen CE, Dover GJ, Kazazian HH (1984) Origin of the ß-globin gene in blacks: the contribution of recurrent mutation or gene conversion or both. Proc Natl Acad Sci USA 81:853-856

Botstein D, White RL, Skolnick M, Davis RW (1980) Construction of a genetic linkage map in man using restriction fragment length polymorphisms. Am J Hum Genet 32:314-331

Brett P, Curtis D, Gourdie A, Schnieden V, Jackson G, Holmes D, Robertson M, Gurling H (1990) Possible linkage of Tourette syndrome to markers on short arm of chromosome 3 (C3p21-14). Lancet 336:1076

Daiger SP, Chakraborty R, Reed L, Fekete G, Schuler D, Berenssi G, Nasz I, Brdicka R, Kamaryt J, Pijáckovȧ A, Moore S, Sullivan S, Woo LC (1989) Polymorphic DNA haplotypes at the phenylalanine hydroxylase (PAH) locus in European families with phenylketonuria (PKU). Am J Hum Genet 45:310-318

Egeland JA, Gerhard DS, Pausl DL, Sussex JN, Kidd KK, Allen CR, Hostetter AM, Housman DE (1987) Bipolar affective disorders linked to DNA markers on chromosome 11. Nature 325:783-787

Gelernter J, Pakstis AJ, Pauls DL, Kurlan R, Gancher ST, Civelli O, Grandy D, Kidd KK (1990) Tourette syndrome is not linked to D_2 dopamine receptor. Arch Gen Psychiatry 47:1073-1077

Goate A, Chartier-Harlin M-C, Mullan M, Brown J, Crawford F, Fidani L, Giuffra L, Haynes A, Irving N, James L, Mant R, Newton P, Rooke K, Roques P, Talbot C, Pericak-Vance M, Roses A, Williamson R, Rossor M, Owen M, Hardy J (1991) Segregation of a missense mutation in the amyloid precursor protein gene with familial Alzheimer's disease. Nature 349:704-706

Hauge XY, Grandy DK, Eubanks JH, Evans GA, Civelli O, Litt M (1991) Detection and characterization of additional DNA polymorphisms in the dopamine D2 receptor gene. Genomics 10:527-530

Heutink P, Sandkuyl LA, van de Wetering, Oostra BA, Weber J, Wilkie P, Devor EJ, Pakstis AJ, Pauls D, Kidd KK (1991) Linkage and Tourette syndrome. (Letter to the Editor) Lancet 337:122-123

Kan YW, Dozy AM (1978) Polymorphism of DNA sequence adjacent to human ß-globin structural gene: relationship to sickle mutation. Proc Natl Acad Sci USA 75:5631-5635

Kelsoe JR, Ginns EI, Egeland JA, Gerhard DS, Goldstein AM, Bale SJ, Pauls DL, Long RT, Kidd KK, Conte G, Housman DE, Paul SM (1989) Reevaluation of the linkage relationship between chromosome 11p loci and the gene for bipolar affective disorder in the Old Order Amish. Nature 342:238-243

Kerem B, Rommens JM, Buchanan JA, Markiewicz D, Cox TK, Chakravarti A, Buchwald M, Tsui LC (1989) Identification of the cystic fibrosis gene: genetic analysis. Science 245:1073-1080

Kidd KK (1987) Phenylketonuria: the population genetics of a disease. Nature 327:282

Kidd KK (1991) Progress towards completing the human linkage map. Curr Opin Genet Devel 1/1:99-105

Kidd KK, Ott J (1984) Power and sample size in linkage studies. HGM7: Seventh International Workshop on Human Gene Mapping. Cytogenet Cell Genet 37:510-511

Kidd KK, Bowcock AM, Schmidtke J, Track RK, Ricciuti F, Hutchings G, Bale A, Pearson P, Willard HF, with help from Gelernter J, Giuffra L, and Kubzdela K (1989) Report of the DNA committee and catalogs of cloned and mapped genes and DNA polymorphisms. HGM10, Cytogenet Cell Genet 51:622-947

Kruger SD, Turner WJ, Kidd KK (1982) The effects of requisite assumptions on linkage analyses of manic-depressive illness with HLA. Biol Psychiatry 17:1081-1089

Lathrop GM, Lalouel JM, Julier C, Ott J (1984) Strategies for multilocus linkage analysis in humans. Proc Natl Acad Sci USA 81:3443-3446

Moises HW, Gelernter J, Giuffra LA, Zarconi V, Civelli O, Wetterberg L, Kidd KK, Cavalli-Sforza LL (1991) No linkage between D2-dopamine receptor gene region and schizophrenia. Arch Gen Psychiatry 48:643-647

Morton NE (1955) Sequential tests for the detection of linkage. Am J Hum Genet 7:277-318

Morton LA, Kidd KK (1981) The effects of variable age-of-onset and diagnostic cirteria on the estimates of linkage: an example using manic-depressive illness and color blindness. Soc Biol 27:1-10

Oberlé I, Rousseau F, Heitz D, Kretz C, Devys D, Hanauer A, Boué J, Bertheas MF, Mandel JL (1991) Instability of a 550-base pair DNA segment and abnormal methylation in fragile X syndrome. Science 252:1097-1102

Pakstis AJ, Heutink P, Pauls DL, Kurlan R, van de Wetering BJM, Leckman JF, Sandkuyl LA, Kidd JR, Breedveld GJ, Castiglione CM, Weber J, Sparkes RS, Cohen DJ, Kidd KK, Oostra BA (1991a) Progress in the search for genetic linkage with Tourette syndrome: an exclusion map covering more than 50% of the genome. Am J Hum Genet 48:281-294

Pakstis AJ, Kidd JR, Castiglione CM, Kidd KK (1991b) Status of the search for a major genetic locus for affective disorder in the Old Order Amish. Human Genet 87:475-483

Riordan JR, Rommens JM, Kerem B, Alon N, Rozmahel R, Grzelczak Z, Zielenski J, Lok S, Plavsic N, Chou JL, Drumm ML, Iannuzzi C, Collins FS, Tsui LC (1989) Identification of the cystic fibrosis gene: cloning and characterization of complementary DNA. Science 245:1066-1073

Risch N, Giuffra L (1991) Model misspecification and multipoint linkage analysis. Hum Hered (in press)

Rommens JM, Iannuzzi MC, Kerem B, Drumm ML, Melmer G, Dean M, Rozmahel R, Cole JL, Kennedy D, Hidaka N, Zsiga M, Buchwald M, Riordan JR, Tsui LC, Collins FS (1989) Identification of the cystic fibrosis gene: chromosome walking and jumping. Science 245:1059-1065

Sarkar G, Kapelner S, Grandy DK, Marchionni M, Civelli O, Sobell J, Heston L, Sommer SS (1991) Direct sequencing of the dopamine D2 receptor (DRD2) in schizophrenics reveals three polymorphisms but no structural change in the receptor. Genomics 11:8-14
Sutcliffe JG (1988) mRNA in the mammalian central nervous system. Annu Rev Neurosci 11:157-198
Yu S, Pritchard M, Kremer E, Lynch M, Nancarrow J, Baker E, Holman K, Mulley JC, Warren ST, Schlessinger D, Sutherland GR, Richards RI (1991) Fragile X genotype characterized by an unstable region of DNA. Science 252:1179-1181

Discussion of Presentation K. K. Kidd

Vogel

I could not agree more to all your comments on associations. However, you told us that you excluded 85% of the genome for linkage with Tourette syndrome and I predict that sooner or later you will also exclude the remaining 15%. The reason is that the Tourette phenotype is extremely weak and therefore the modes of inheritance are very spurious. The vagueness of this phenotype turns out from the prevalence figures. The prevalence figures of the Tourette syndrome phenotype and the American population range, depending on the authors, between 1 in 100 000 and 1 in 100. There might be a core of real Tourette cases, but the Tourette spectrum is of course very wide.

Kidd

I absolutely agree that there is a great deal of discussion and controversy over the Tourette diagnosis. I am a member of a group at Yale that represents the opposite pole from David Commings, who says that everything is caused by the same gene, including all of the disorders we have been talking about here today. Most of our analyses at Yale are done using a very rigorously defined and quite narrow core phenotype. We have put a lot of effort into the genetic epidemiology and we do not find much ambiguity in that core phenotype. I hasten to add that I would not be too surprised if we do reach a 100%. But if we have to have an epistatic model, or a model of genetic heterogeneity, our studies are appropriately designed. We are part of an international consortium; we send the DNA of our big pedigrees to other labaratories to type their markers and they send theirs to us. We are going to end up with a set of pedigrees, each one of which has 300–400 markers on it, spanning the genome, and if TS is more complex, then this is exactly the type of data set needed to try to look at heterogeneity or epistatic models pedigree by pedigree.

McGuffin

I of course completely disagree with Dr. Vogel about association, but I completely agree with him about his comments on Tourette syndrome. Let me say something more about the Tourette syndrome problem. We probably will end up excluding the entire genome. But one of the problems is that in developing enthusiasm for linkage studies in Tourette, you might have overlooked one of the basic problems of psychiatric genetics. The most efficient way to demonstrate that there is really a genetic basis as opposed to familial basis for a disorder is to carry out the good old-fashioned twin and adoption studies. There have been twin studies in Tourette syndrome, but they have not been carried out with the same vigor and thoroughness as studies in affective disorder and schizophrenia. So it could well be that we will end up doing a systematic search for schizophrenia and our conclusion will be there is no major gene for schizophrenia, but there are genes somewhere. The trouble with Tourette syndrome may be that you end up excluding the entire genome not knowing whether you missed a gene or whether there was a gene or any genes at all there in the first place.

Kidd
Well, I disagree slightly. There have been some adoption studies done, not systematic adoption records, but simply based on adoptees that have shown up in the families and thus have come to attention. In no case has Tourette syndrome been transmitted across the social connection that is not a biological connection. Among several probands who were adopted, none had a positive family history among their nonbiological "relatives," whereas 90% of the probands have biological relatives who are affected. And conversely, some of our probands have had adopted children and in no cases have we found an adopted child who had Tourette syndrome. I also disagree that twin studies are one of the best ways to show familiality. I do not think they are any better than the classical family studies; they suffer from most of the same problems. The adoption study is really the only definitive way to exclude any sort of environmental phenomenon. Certainly, by a variety of types of analyses in systematically ascertained families, Tourette syndrome has very high familiality, quite consistent with a single locus.

McGuffin
You are probably aware of our papers on replication of the recessive gene for attending medical school that was based on a classical family study method and analyzed using a most outstanding segregation analysis. It was a replication study, because 30 years earlier using a more simple method of segregation analysis, the same result was found. So, if you take classical family studies on their own, they can be very misleading. I am not saying that you have to all focus on twin or adoption studies, you have got to have the full pack, the family, the adoption, and the twin studies.

Kidd
The twin studies that we have been involved in are clearly not systematic, but MZ twins have much higher concordance than DZ twins. These findings are quite consistent with the same model of inheritance that we find. The samples are not very large, but I find the data quite solid. I certainly agree with the point that many things can simulate genes. I have been working on Tourette syndrome for 13 years now and have worked on bipolar disorders for 17 years and schizophrenia for close to 20 years. I find the evidence for single locus inheritance much stronger for Tourette syndrome than for either of the others. But, I hasten to add, it is not proven. Nevertheless there is pretty strong evidence for single-locus inheritance.

Gurling

I wonder if we are talking about Tourette syndrome as a test case. We are really sort of projecting our fears onto this illness. We are really talking about schizophrenia and if we were to do the whole genome in schizophrenia and manic depression and find absolutely nothing, it does not really matter. Then we can move on to the next hypothesis, and all the preexisting data will still be as useful as it was to exclude the genome and the single-locus models.

Barnard

I would like to say something about the stricture of our attending the genes. I think there are arguments for using candidate genes, where possible. But, the converse is also true, where one actually knows what the gene is. It is not immediately going to reveal to us the etiology and one can point of course to well-known long-established cases, where the precise gene is known for a psychiatric defect. With Tay Sachs or Lesh-Nyhan, the models come to mind where we know precisely what the enzyme defect is, but it does not tell us anything about the causation of the complex symptomatology. So, it is very complicated to go from the immediate cause to the pathology and that is an argument which works in both directions.

Kidd

I absolutely agree and I made my statement on candidate genes very strong simply to make the point. But, there are contrary examples and as I said, I am also studying some candidate genes. A very good recent example is the case of Alzheimer's disease in which there was a candidate gene which for reasons not yet completely clear, genetic heterogeneity, misdiagnosis or something, was thought to be excluded, just because of a couple of obligate crossovers. The linkage data indicated the general region, but excluded this candidate locus. Only by going back and actually sequencing the gene in affected individuals did they find that it is the etiologically relevant locus. So one of the problems with all of these studies is not just the false positives, but the false negatives as well. The kind of simulation study I was mentioning to look at the stability of a positive lod score can equally be done conversely by taking a large group of negative data and systematically removing one affected individual at a time to find out whether the diagnosis in one individual might be the primary contributor to the exclusion of that region. This is post hoc and has to be recognized as exploratory data analysis, hypothesis generation. But it nonetheless might allow one to focus on critical individuals, on diagnoses that might be most suspect.

Mendlewicz

Could you comment on the issue of the unaffected subjects in the pedigree. How would you best deal with that? Would you, in first analyses, simply delete them and only analyze the affected ones? Do you think this might be the right way?

Kidd

One of the problems in doing that is that there is often confusion over what is to be deleted. I might delete their diagnoses, but I would not delete their marker types. I would leave the individuals and their marker data in the pedigree, because they contribute a tremendous amount to the inference of how the chromosomes have actually been transmitted within the family and replace the diagnosis of "unaffected" with "unknown." Unknown phenotype is virtually analogous to the penetrance free model. That is certainly one possibility. I tend to leave them in as unaffected for the initial analyses since I look at all of these analyses as model and assumption dependent. But, when one wants to get more rigorous, one may find that a positive result is entirely dependent on the unaffected individuals, which is another reason for them being suspect.

Beyreuther

As to Dr. Barnard's and Dr. Kidd's mention of knowing a locus and knowing a gene, even if we do know a gene but do not know the consequences of the mutation that we have localized, it is very important to have the definitive diagnosis. If you have a gene, you can in principle do a systematic study, regarding drug design with a single patient. I think that is the very important issue, e.g., in Alzheimer's disease, where you otherwise have to do a study with a large cohort, with a lot of false-positive ones. In future that will be the approach and what the Swedes did with the Parkinson's patients, showing us that a very complicated technique such as PET can provide much information about the disease.

Reich

Many of us are interested in the range of plausible models which we may obtain when we study any functional disorder. The rationale for that is the design of any study, whether they are explicit or implicit depends to some extent on what we think about transmission. And since we do not know, I think it is important to determine the range of plausible models which could be studied by the means we are using. In that regard, I think there is no right or wrong way; there is only the range of influences one can test within a given context. I believe probably the most flexible strategy, the one that is best buffered against uncertainity, is one that includes both large and small families, one that enables affected sib-pair methods as well as single large extended dense pedigrees to be studied simultaneously. Hopefully, if measurements are sufficiently more standardized, we will be able to test whether we find the same thing and whether we are looking for it in the same way. So, from my point of view, a multifront strategy with different types of studies is the optimal design for assuring success, since the mode of transmission is simply unknown.

Van Broeckhoven

I just wanted to remind Dr. Kidd of the fact that the mutation in Alzheimer's disease has been found in other families and that the findings of families that you quote as so-called obligate recombinants were still found in two families that show linkage to chromosome 21 and don't show the mutation. And the candidate gene analysis, which relates directly to the APP gene on chromosome 21, is another example that a candidate gene approach can be successful.

Clinical Aspects of Familial Forms of Alzheimer's Disease

L. Amaducci, P. Forleo, P. Piersanti, and S. Sorbi

Familial Alzheimer's disease (FAD) has been recognized at least since Lua's report in 1920. Since then over 100 families with FAD have been reported in different countries.[1] Even though nongenetic cases may possibly occur, it is unlikely that the aggregation of affected in families can be explained by chance alone. Autosomal dominant inheritance with age-dependent penetrance has been suggested to explain this aggregation in most of these families. At least 70 reported families contain affected members in at least 3 generations, suggesting an autosomal dominant mode of inheritance in these families. The reports from the remaining families do not contain sufficient information to differentiate between a commonly occurring recessive gene, a dominant transmitted disorder, or multiplex clustering of a nongenetic trait. Analysis of pedigrees segregating a disease phenotype may suggest possible modes of transmission. In addition such pedigrees may allow detection of a unique feature which defines specific subtypes of the disease.

However, the majority of published reports of families with FAD are from pedigrees selected according to different criteria from originally larger groups of families. This has, indeed, been the rule since the appearance of molecular biology and the application of linkage analysis techniques which require the utilization of pedigrees with defined and homogeneous modality of transmission. Because of that the pedigrees reported so far are probably not representative of all families with FAD but represent a sample with the bias of homogeneity. Nevertheless, the analysis of the published reports of these families suggests some phenotypic heterogeneity in familial Alzheimer's disease.

---

[1]See the following references: Lowenberg and Rothschild (1931); Schottky (1932); Von Braunmühl (1932); James (1933); Lowenberg and Waggoner (1934); Rothschild (1934); Franc (1936); Grunthal and Wenger (1939); McMenency et al. (1939); Van Bogaert et al. (1940); English (1942); Delay et al. (1945); Essen-Moller (1946); Luers (1947); Gosch (1949); Arndt (1951); Fattovich (1952); Larsson et al. (1953); Neuman and Cohn (1953); Corsellis and Brierky (1954); Marchand et al. (1954, 1956); Josephy (1956); Von Braunmühl (1959); Wheelan (1959); Zawuski (1960); Lauter (1961); Hirano and Zimmermann (1962); Friede and Magee (1962); Feldman et al. (1963); Nahman and Rabinowicz (1963); Bucci (1963); Jelgersma (1964); Costantinidis et al. (1965); Heston et al. (1966); Maeda et al. (1968); Landy and Bain (1970); Jacob (1970); Beigthon and Lindenberg (1971); Cook et al. (1979); Goudsmit et al. (1981); Nee et al. (1983;) Powell and Folstein (1984); Ebinger et al. (1987); Gimenez-Roldan et al. (1988); Sadovnick et al. (1988); Bird et al. (1989).

FAD Pedigrees

Flugel first reported in 1929 one patient from a family with three affected in two generations. Since then case reports of 47 families have been published in different papers[2]; 3 papers report 2 families each (Van Bogaert et al. 1940; Fattovich 1952; Goudsmit et al. 1981), 2 papers report 3 families each (Cook et al. 1979; Larsson et al. 1953), 1 paper reports 4 pedigrees (Powell and Folstein 1984), 1 paper reports 11 families (Constantinidis et al. 1965), 1 paper (Bird et al. 1989) reports 24 families with autosomal dominant transmission selected from a group of more than 80 families; and a recent paper (Farrer et al. 1990) reports 70 families with autosomal dominant transmission collected by the international FAD group and selected from larger groups of pedigrees to fulfill some criteria for a specific linkage study. Some of the features reported in these papers are summarized in Table 1.

Table 1. Some features of these 149 families

	Affected (*n*)	Neuropathology confirmed (*n*)	Affected/ total in sibship (*n*)	Percentage affected in sibship	Age at onset Mean ± SD (range)
Case reports (papers with 1-4 families)	366	69	188/348[a]	54	41.45 ± 14 (25-89)
11 Families	27	17	27/55	49	-
24 Families	151	61	151/347	44	54.7 ± 11.5 (30-84)
70 Families	495	197	495/906	55	60.9 (35-80)
149 Pedigrees	1039	344	861/1656	52	53 (25-89)

[a]Calculation possible only in 188 cases from the total of 366 affected described.

[2]The pertinent references can be found in footnote 1, specifically, those which have not otherwise been cited in this paragraph.

Clinical Features

Type of Segregation

In over 100 pedigrees 52% of adult members developed dementia, a proportion which approximates the 50% segregation ratio expected for an autosomal dominant trait with complete penetrance. This observed ratio, however, does not consider biases introduced either by the manner of ascertainment or by failure to consider the ages of unaffected members relative to the age at onset in affected members. In fact only some, and not all, pedigrees contain sufficient numbers of affected individuals in multiple generations to provide a rigorous argument for a specific modality of inheritance for that specific family, namely, a commonly occurring recessive gene, or autosomal dominant, or a clustering of a nongenetic trait.

Age at Onset

Apart from age at onset, only few different phenotytpes are apparent from these published reports of the pedigrees. Families in which the affected are younger at onset are predominant in these studies, although several late onset cases have been described. This observation may indicate that "genetic" AD is more likely to have an early onset or simply reflects the fact that early onset pedigrees are more easily identified and studied.

It is a matter of fact that the age at onset was reported to be below 60 years in all families until Powell and Folstein in 1984 described pedigrees in which age at onset varied between 25 and 89 years. In a number of these families (Flugel 1929; see also references cited in footnote1) age at onset was under 40 years.

The few more recent publications of series of families with AD (Bird et al. 1989; Farrer et al. 1990) report both early and late onset.

Bird and colleagues described in 1989 the clinical and pathological characteristics of 24 kindreds from a group of 80 families with a history of dementia. This probably represents the largest collection of families with pathologically confirmed FAD compared by a single group of investigators. As in the few other series available, the families were not randomly ascertained and therefore the statistics do not necessarily represent the frequency of various possible types of FAD in the general population. Mean age at onset was 56 + 12 years with a very wide range of 30 to 84 years. This large difference in age at onset was not solely dependent on the difference between families. In fact, for example, in family KH in the series of Bird et al. (1989) a son had become demented at age 43, his mother years later when she reached age 84. Duration of the disease in all of these pedigrees was, however, quite similar and constant (8 ± 4 years). A special effort to test heterogeneity was recently made by Farrer and colleagues (1990), who studied 70 families with FAD, the data from some having already been published (Nee et al. 1983; Bird et al. 1989; Foncin et al. 1985). Unfortunately only few clinical features of these pedigrees are available.

However, Farrer et al. (1990) reported that a familial mean age at onset of 58 years was the most appropriate breakpoint between early and late onset families. Lifetime risk of AD among first-degree relatives of an early onset AD cohort was significantly less than among relatives of late onset cohorts (53% versus 87%). In late onset families there was also a higher risk among women than men (68% versus 54%). Thus, this study suggests heterogeneity for FAD. The lifetime risk of 53% observed in the early onset family is consistent with an autosomal dominant model. A single allele cannot account alone for the higher (86%) risk to offspring of affected parents in late onset pedigrees and for predominancy of the disease in women.

However, Mayeux et al. (1985) studying 121 unrelated consecutive nonfamilial cases of AD found onset age distributed bimodally but noted no clinical differences between early and late onset.

Duration of the Disease

The duration of the disease was shorter only for those who died before 1950 (6 versus 9 years). The overall male to female ratio was 1.1:1.0. As in all other single families or series of FAD pedigrees reported so far, 51% of individuals at risk developed the disease if they lived beyond the mean age at onset of the family. Only five of the families reported in this study had an early onset (mean of 41 years). While the duration of the disease was not different from that observed in late onset families, other clinical features were dissimilar.

Symptoms and Signs

Major motor seizures occurred significantly more frequently in early onset disease, being present in 33% of patients in early onset versus 3% of patients of late onset pedigrees. Language deficits were present in 50% versus 32%, muscle rigidity in 33% versus 21%, tremor in 17% versus 9%, and myoclonus in 19% versus 6% in the large series described by Bird et al. (1989). It is difficult to compare this carefully clinically examined series of pedigrees with the others published because most papers do not report the presence or the absence of these clinical signs. However, seizure and myoclonus in FAD patients have been reported in a number of other pedigrees of early onset including the very large one described by Foncin et al. (1985), the probably genetically related one described by Feldman et al. (1963), the pedigree of Nee et al. (1983), and one of the two pedigrees of Martin et al. (1991), while these clinical signs are rarely reported in late onset families (onset age after 65 years).

Clinical amyotrophy and anterior cell loss at autopsy have occasionally been reported in affected members of families with early onset FAD (one member of FAD, see St. George-Hyslop et al. 1987), with late onset FAD (Bird et al. 1989), sporadic cases (Thomas et al. 1982) and in nondemented members of families with early onset FAD (Goudsmit et al. 1981). These findings are of particular interest both for the possible localization of a gene causing FAD in chromosome 21 (St. George-Hyslop et al. 1987) and for the recent possible localization of the gene causing familial amyotrophic lateral sclerosis (ALS) using polymorphic markers mapping on the same chromosome (Siddique et al. 1991). It will be of interest to examine the markers linked to a possible ALS gene in chromosome 21 in those families with familial and/or sporadic AD in which ALS or an alteration resembling ALS in demented subjects have been reported. These families may prove to represent a particular variant or a subgroup of Alzheimer's dementia.

Schizophrenialike onset progressing to severe dementia is another clinical feature reported occasionally in some members of either early or late onset FAD. It is not clear whether the presence of psychotic signs as predominant signs at onset reflects a variant of FAD or a distinct entity. However, in at least one member of such a family (Bird et al. 1989) duration of the disease was unusually long (mean 11 years, but up to 20 years), and autopsy demonstrated excessive neurofibrillary tangles (NFT) in the brain amygdala, temporal cortex, nucleus basalis of Meynert, and dorsal raphe nucleus but not in the hippocampus. Neither neuritic or amyloid plaques nor amyloid angiopathy was found in these brains, suggesting that at least this family may suffer from a variant of FAD, despite the clinical features of AD-like dementia and the presence of NFT.

Down's syndrome has been reported in a few families with both early and late onset FAD (Bird et al. 1989; St. George-Hyslop et al. 1987).

Ethnic Differences

A peculiar set of pedigrees was described by Bird and colleagues (1988) — descendants of the Volga Germans. While the mean age is slightly higher (55 years, range 40-71) than is generally the case in the other published reports of families with early onset the clinical characteristics were not different, including the presence of myoclonic jerking. At least one member of one of these Volga pedigrees (family E) had spongiform changes in the cerebellum and hippocampus, suggesting Creutzfeldt-Jakob disease. It is interesting to note that while Bird and colleagues (1988) are proposing that those families may share a single affected common ancestor, or genetic founder, thus representing one large pedigree with a single dominant mutation, the clinical picture is similar to that of the other families with early onset FAD according to reports published so far.

Neuropathology

No systematic data are available comparing the neuropathology of different FAD families. As reported above, in the families with schizophrenialike onset there is a major neuropathological peculiarity, namely, the absence of neuritic or amyloid plaques (Bird et al. 1989).

A recent paper suggests that a subgroup of FAD families may be represented by those having a mutation in the amyloid gene (Lucotte et al. 1991; Goate et al. 1989, 1991). It is particularly interesting that some of these individuals with such mutation have amyloid angiopathy (Lucotte et al. 1991). It is already known that a mutation in the amyloid precursor protein gene is also the site of the mutation leading to hereditary cerebral hemorrhage with Dutch-type amyloidosis (Levy et al. 1990).

Conclusions

In conclusion, a number of papers report pedigrees in which AD affects several members in several generations. The families which are best described have early onset FAD. This probably reflects the method of ascertainment of families with FAD, which usually requires the presence of pathologically confirmed cases and several affected in different generations. However, because the age at onset may vary consistently within the same family, of course it happens more often in early than in late onset families that offspring live well into their individual family's risk period. Thus, families with late onset FAD are more difficult to discover and, if so, they may represent as suggested by Farrer et al. (1990) a carefully, though inadvertently, selected set of families which have a pattern of familial aggregation that is atypically intense.

On the basis of published pedigrees, with the limitation of a number of biases possibly present, there is some evidence of clinical heterogeneity in FAD families. The most important phenotypic feature seems to be age at onset, which is consistent within a family and which allows at least differentiation of early and late onset FAD pedigrees.

In early onset FAD families, while duration of the disease after 1950 is no different from that found in late onset families, some other clinical features are constantly more frequent. These include major motor seizure, myoclonic jerking, and rigidity. In some individuals with dementia and members of FAD pedigrees the possibility of a spongiform encephalopathy has been suggested by specific neuropathological findings but ruled out in the majority of cases at autopsy. A reexamination of these cases by the recent polymerase chain reaction techniques for prion mutation might provide further useful information.

Finally, despite the fact that psychological changes are common aspects of the neurobehavioral changes of AD, there are reports of pedigrees in which members are affected by schizophrenia before the appearance of memory and other cognitive changes. In some of these families this may be a result of a variant of FAD or a different disorder.

Heterogeneity is plausible in a disease as complex as AD, as is the case in other neurological disorders such as Charcot-Marie-Tooth atrophy, olivopontocerebellar atrophy, torsion dystonia, and familial amyloid polyneuropathy, which have multiple genetic etiologies but are clinically indistinguishable.

References

Arndt M (1951) La form familial de la enfermedad de Alzheimer. Neuropsiquiatria 2:169-181

Beigthon PH, Lindenberg R (1971) Alzheimer's disease in multiple members of a family. Birth Defects 7(VI):232-233

Bird TD, Lampe TH, Sumi SM et al. (1988) Familial Alzheimer's disease in American descendants of the Volga Germans. Probable genetic founder effect. Ann Neurol 23:25-31

Bird TD, Sumi SM, Nemens EJ et al. (1989) Phenotypic heterogeneity in familial Alzheimer's disease: a study of 24 kindreds. Ann Neurol 25:12-25

Bucci LA (1963) A familial psychosis of Alzheimer type in 6 kinships of 3 generations. Am J Psychiatry 63:863-866

Cook RH, Ward BE, Austin JH (1979) Studies in aging of the brain: V. Familial Alzheimer's disease: relation to transmissible dementia, aneuploidy and microtubular defects. Neurology 29:1402-1412

Corsellis JAN, Brierley JB (1954) An unusual type of presenile dementia. Brain 77:571-587

Costantinidis J, Garrone G, Tissot R (1965) L'incidence familiale des lesions neurofibrillaires d'Alzheimer corticales. Psychiatra Neurol 150:234-247

Delay J, Deselaux P, Perrin J, De Burat JF (1945) Demence atrophique familiale. Aspects encephalographiques analogues de deux collateraux. Rev Neurol (Paris) 77:85

Ebinger G, Bruyland M, Martin JJ et al (1987) Distribution of biogenic amines and their catabolites in brains from patients with Alzheimer's disease. J Neurol Sci 77:267-283

English WH (1942) Alzheimer's disease. Psychiatr Q 16:91-106

Essen-Moller E (1946) A family with Alzheimer's disease. Acta Psychiatr Scand 21:233-244

Farrer LA, Myers RH, Cupples LA et al. (1990) Transmission and age-at-onset patterns in familial Alzheimer's disease. Evidence for heterogeneity. Neurology 40:395-403

Fattovich G (1952) La maladie d'Alzheimer-Perusini a caractere heredo-familial. Rass Studi Psichiatr 12: 895

Feldman RG, Chandler KA, Levy LL (1963) Familial Alzheimer's disease. Neurology 13:811-823

Flugel FE (1929) Zur Diagnostik der Alzheimerschen Krankeit. Z Gesamte Neurol Psychiatr 120:783-787

Foncin JF, Salmon D, Supino-Viterbo V et al. (1985) Demence presenile d'Alzheimer transmise dans une famille etendue. Rev Neurol (Paris) 141:194-202

Franc DB (1936) Die familiare Form der Alzheimerschen Krankheit. Sov Psychoneurol 12:15

Friede RL, Magee KR (1962) Alzheimer's disease. Presentation of a case with pathologic and enzymatic histochemical observations. Neurology 2:213-222

Gimenez-Roldan S, Mateo D, Escalona Zapata J (1988) Familial Alzheimer's disease presenting as levodopa responsive Parkinsonism. Adv Neurol 45:431-436

Goate AM, Haynes AR, Owen MJ et al. (1989) Predisposing locus for Alzheimer's disease on chromosome 21. Lancet 1:352-355

Goate AM, Chartier-Harlin MC, Mullan M et al. (1991) Segregation of a missense mutation in the amyloid precursor protein gene with familial Alzheimer's disease. Nature 349:704-706

Gosch H (1949) Alzheimer's Krankheit und prosektische Katatonie. Psychiatr Neurol Med Psychol (Leipz) 1:302

Goudsmit J, White BJ, Weitkamp LR et al. (1981) Familial Alzheimer's disease in two kindreds of the same geographic and ethnic origin. J Neurol Sci 49:79-89

Grunthal E, Wenger O (1939) Nachweis von Erblichkeit der Alzheimerschen Krankheit nebst Bemerkungen über den Altervorgang im Gehirn. Monatsschr Psychiatr Neurol 101:8

Heston LL, Lowter DL, Lenethol CM (1966) Alzheimer's disease: a family study. Arch Neurol 15:225-233

Hirano A, Zimmermann HM (1962) Alzheimer's neurofibrillary changes. Arch Neurol 7:227-242

Jacob H (1970) Muscular twitchings in Alzheimer's disease. In: Wolstenholme GEW, O'Connor M (eds) Alzheimer's disease and related conditions. Churchill, London, pp 75-93

James GWB (1933) Discussion on the mental and physical symptoms of the presenile dementias. Proc R Soc Lond [Biol] 26:1088

Jelgersma HC (1964) Ein Fall von juveniler hereditärer Demenz vom Alzheimer Typ mit Parkinsonismus und Kluver-Bucy Syndrom. Arch Psychiatr Nervenkr 205:262-266

Josephy H (1956) A clinicopathologic report of Alzheimer's disease. J Neuropathol Exp Neurol 15:234

Landy PJ, Bain BJ (1970) Alzheimer's disease in siblings. Med J Aust 2:832-834

Larsson R, Sjogren T, Jacobson G (1953) Senile dementia: a clinical sociomedical and genetic study. Acta Psychiatr Scand 39 [Suppl]:1-259
Lauter H (1961) Genealogische Erbebungen in einer Familie mit Alzheimerschen Krankheit. Z Gesamte Neurol Psychiatr 202:126-139
Levy E, Carman MD, Fernandez-Madrid IJ et al. (1990) Mutation of the Alzheimer's disease amyloid gene in hereditary cerebral haemorrhage, Dutch type. Science 248:1124-1128
Lowenberg K, Rothschild D (1931) Alzheimer disease. Its occurrence on the basis of a variety of factors. Am J Psychiatry 37:280-284
Lowenberg K, Waggoner RW (1934) Familial organic psychosis Alzheimer's type. Arch Neurol Psychiatry 31:737-754
Lua M (1920) Zur Kasuistik der Alzheimerschen Krankheit. Z. Gesamte Neurol Psychiatry 55:60
Lucotte G, Berriche S, David F (1991) Alzheimer's mutation. Nature 351:438
Luers T (1947) Über die familiäre juvenile Form der Alzheimerschen Krankheit mit neurologischen Herderscheinungen. Arch Psychiatr Nervenkr 179:132-145
Maeda S, Aratame G, Osanai Y et al. (1968) A case of familial Alzheimer's disease. Psychiatr Neurol Japan 70:419-428
Marchand L, Sivadon P, Koechlin P et al. (1954) Maladie d'Alzheimer ayante debute a 38 ans. Hereditye chargee. Soc Med Psich Mai 10:120-128
Marchand L, Le Conte M, Halpern B (1956) Considerations pathogeniques sur les dimenies atrophiques. Presse Med 64:1412
Martin JJ, Ghuens J, Bruyland M et al. (1991) Early onset Alzheimer's disease in two large Belgian families. Neurology 41:62-68
Mayeux R, Stern Y, Spanton S (1985) Heterogeneity in dementia of the Alzheimer type: evidence of subgroups. Neurology 35:453-461
McMenency G, Worster-Draught C, Flind J, Williams HG (1939) Familial presenile dementia: report of case with clinical and pathological features of Alzheimer's disease. J Neurol Psychiatry 2:293
Nahman S, Rabinowicz T (1963) Considerations sur un cas familial de maladie d'Alzheimer. Encephale 52: 369-381
Nee L, Polinsky RJ, Eldridge R et al. (1983) A family with histologically confirmed Alzheimer disease. Arch Neurol 40:203-208
Neuman MA, Cohn RC (1953) Incidence of Alzheimer's disease in a large mental hospital. Arch Neurol Psychiatry 69:615-636
Powell D, Folstein MF (1984) Pedigree study of familial Alzheimer disease. J Neurogenet 1:189-197
Rothschild D (1934) Alzheimer's disease. A clinicopathologic study of 5 cases. Am J Psychiatry 91:485-519
Sadovnick AD et al (1988) Familial Alzheimer's disease. Can J Neurol Sci 15:142-146
Schottky J (1932) Über praesenile Verblödungen. Z Gesamte Neurol Psychiatry 140:333-397
Siddique T, Figlewicz DA, Pericak-Vance M et al. (1991) Linkage of a gene causing familial amyotrophic lateral sclerosis to chromosome 21 and evidence of genetic locus heterogeneity. N Engl J Med 324: 1381-1384
St. George-Hyslop PH, Tanzi R, Polinsky R et al. (1987) The genetic defect causing familial Alzheimer's disease maps on chromosome 21. Science 235:885-890
St. George Hyslop PH (1989) Familial Alzheimer's disease: progress and problems. Neurobiol Aging 10: 417-425
Thomas M, Ballantyne JP, Hansen S et al. (1982) Anterior horn cell dysfunction in Alzheimer disease. J Neurol Neurosurg Psychol 45:378-381
Van Bogaert L, Maere M, De Smedt E (1940) Sur les formes familiales precoces de la maladie d'Alzheimer. Monatsschr Psychiatr Neurol 102:249-301
Von Braunmühl A (1932) Kolloid chemische Betrachtungsweise seniler und praeseniler Gewebsveränderungen. Z Gesamte Neurol Psychiatr 142:4
Von Braunmühl A (1957) Alzheimersche Krankheit. In: Scholz W (ed) Erkrankungen des zentralen Nervensystems. Springer, Berlin Göttingen Heidelberg New York, pp 500-501 (Handbuch der speziellen pathologischen Anatomie, vol 13, part 1)
Wheelan L (1959) Familial Alzheimer's disease. Ann Hum Genet 23:300-310
Zawuski G (1960) Zur Erblichkeit der Alzheimerschen Krankheit. Arch Psychiatr Neurol 201:123

Discussion of Presentation S. Sorbi/L. Amaducci et al.

Crow

There have been some cases in the literature of familial Alzheimer's disease which have been rediagnosed as prion related after further investigations. The pedigree from Indiana was started like that. How thoroughly can prion-related disorders be ruled out in the ones that you are describing now?

Sorbi

Several of these families were described between 1920 and 1945, and the only data available are neuropathological descriptions. We have tried to count the affected probands to see the clinical presentation. Probably most of the symptoms and signs which I have reported are not reported in the old papers. In some papers there is just a neuropathological description. They say that this is the major process. Maybe there were also some schizophrenialike symptoms, but they simply did not report this data.

Beyreuther

The first Alzheimer family, identified by Heston, was the Indiana family. This family shows tangles, no Alzheimer plaques, but POP plaques. Now, you mentioned another family having tangles but no other Alzheimer pathology. I think that among those families there was still a considerable number of Alzheimer cases, maybe frontal lobe dementia cases that are familial, frontal lobe with ALS. As far as I remember the ratio now that we got from the Swedish studies was at least one frontal lobe cases in five Alzheimer cases. So there may be many of these that are not real cases of Alzheimer's. And some of these frontal lobe cases do not show any Alzheimer pathology. Is Alzheimer pathology established in the patients from all the families you mentioned here?

Sorbi

All the cases reports are neuropathological case reports. There is quite a good neuropathological description, in about 64 papers. I think they fulfill the criteria for Alzheimer's disease more or less. There are some problems with the few large series of families, because there is little information provided in their papers. It would be desirable to have more data from all the groups studying familial cases to see what the clinical appearance really is, the frequency of the symptoms, and the true neuropathological results. The family without plaques was a family with predominant psychiatric signs, schizophrenialike symptoms, and very long disease duration, about 20 years, so it was probably a strange family for that disease. One of the problems is that most of these families are used less for linkage study — not regarding the differences between families. They are all pooled together.

Hippius
In your slide with the information on the Italian families there was one point of interest for a clinician. Only in two families were there psychotic symptoms and three crosses, three crosses in one family, three crosses in another family, and in all the other five families nothing. What does this mean and do you have information on these pedigrees, on the manifestations of disease, for instance, of schizophrenia? Have you only investigated pedigrees for Alzheimer's disease? What do three crosses mean: very strong, very frequent?

Sorbi
Three crosses mean it was just present at the beginning of the disease. Psychiatric symptoms were often present at first examination of the patients. Actually the two families are probably related. The first one is the Italian family linked to chromosome 21. The second one, with psychotic symptoms, is a family living close to Florence which I found recently. But there is no true evidence that they are originally from the same area. So they probably are connected and are possibly the same family. But we so far have found 16 families with Alzheimer's disease and we presented the data because we have some linkage studies in these seven families. Myoclonus is the latest sign, present in more than 20%–30% of our patients.

Maier
As far as I know, there is a family well known in the literature, the Wetterberg family, which is highly loaded with dementia and schizophrenia. There is a familial aggregation of dementia in that particular family. I do not know what kind of dementia this is, but it would be very useful to know whether this is a dementia which is some how related to Alzheimer's disease. But I have another question. In one of the early reports Heston described the fact that in nonaffected, nondemented relatives of Alzheimer patients there is an increased rate of lymphoblastoma and neoplasms. Is this finding replicated in some other families or is it a peculiar finding in the Indiana family?

Sorbi
I think that these probands have been taken into consideration in a number of case control studies, but so far no positive evidence of increased risk of lymphoblastoma or any other kind of malignancy has been found in first-degree relatives. There are at least five or six case control studies which include first-degree relatives. But it has not been proven to be true. I do not remember any single pedigree except that one reporting a higher incidence of lymphoblastoma.

Mendlewicz
From the clinical point of view I think that Dr. Beyreuther commented that we may possibly have some biological tools to identify some of the individuals at risk, in

whom the disorder is not yet manifest. And in Alzheimer'disease it is late onset illness. In your opinion are there biological markers that could be useful? I think you mentioned brain imaging and PET scanning.

Sorbi
I do not think actually that studies have been performed in the families. Family studies would be best to test that and get more information.

Beyreuther
What I meant yesterday is that there are five families known in the world, but there are more that have a mutation in the amyloid gene and there are already preclinical cases being followed. Preliminary data from PET scanning are very much in support of what you said regarding rigidity and gate disturbances. Patients at risk first show signs of rigidity and gate disturbances. Preclinical work is being done, and PET scanning is being performed. It looks very promising. By following the gate disturbance and rigidity phenomena, there is quite a substantial literature on hip fracture an Alzheimer's disease and the fact is that hip fracture occurs in Alzheimer's disease about five times more often than in age-matched controls. This is done for sporadic Alzheimer's disease. The question is whether there is a similar study also being done for familial Alzheimer's disease?

Sorbi
Not to my knowledge.

Crow
When you speak about the Wetterberg pedigree, are you speaking about the Swedish pedigree that Ken Kidd reported on? Could we get Ken Kidd to tell us more about that family? Is it really common?

Kidd
I am not aware of dementia in that pedigree. I am well aware of an autosomal recessive mental retardation, but that is completely unrelated. It is an ordinary early childhood severe mental retardation of an autosomal recessive sort. To my knowledge it is not anything like a real dementia in the Alzheimer sense and it seems to be completely independent of the schizophrenia in that kindred; it is likely, just as in the Amish, isolates with lots of marriages of third and fourth cousins that are out of some recessive disorders. I have one other question regarding the frequency of Alzheimer's disease pedigrees. You did mention in the paper that the 50% more or less figure across these various published studies is calculated without any ascertainment correction. Did I understand correctly, as this means it is really a highly biased figure?

Sorbi
To reach this value we put all the pedigrees we found in the 64 published papers on the table. We tried to count all the affected and nonaffected sibships; this is where the number came from.

Hippius
From the clinical point of view I find it slightly unsatisfactory that in genetic research very often only the diagnoses of, for example, schizophrenia are used in the investigations. We should remember that the problem of early onset dementia was developed on the basis of dementia paralytica, which was very well known in the last century already. Then the first early onset dementia was the dementia praecox, defined by Kraepelin. And on this basis, 10 years later, Alzheimer, who was a coworker of Kraepelin, defined the next type of early dementia, Alzheimer's disease. We should bear in mind that early onset dementia is not only a problem of the so-called Alzheimer's disease, it is true in other fields. For instance, we have an example at present, HIV early onset dementia or late onset dementia. I don't think it is a matter of nosology from the clinical point of view, and therefore I would like to have such a point of view included in genetic research, too.

Beyreuther
I think in dementia we are faced with the problem of reaching a threshold of brain function. If we are very close to the threshold level, patients may develop all kinds of symptoms that are only due to the threshold problem at the very moment. And if you go beneath the threshold of synaptic density or whatever, it becomes the picture of Alzheimer's disease. I think approaching a threshold might be a general problem for all kinds of diseases. And that was what Kraepelin also has shown.

Barnard
Can I ask for some clarification about the incidence of chromosome 21-linked Alzheimer's disease. Since, as you noted, the cases reported in Britain, the familial cases, are chromosome 21 linked and you said that only one out of the numerous Italian cases seem to be linked to chromosome 21. And is the evidence for nonlinkage completely established, or is it still an open question?

Sorbi
When I said that this was the only family linked to chromosome 21, it is because the other one, which is the London family, is linked to a different portion of chromosome 21. It seems at least more linked to the β-amyloid mutation, but in this family there is definitely no mutation.

Mendlewicz

Dr. Beyreuther spoke about the question of the threshold and when you talk about Alzheimer's disease, you also have to deal with the issue of cognitive dysfunction. As you know, there are many studies on age-related cognitive dysfunction and the potential relationship with onset of Alzheimer's disease. Are you aware of any genetic family studies in the area of age-associated memory impairment syndrome (AAMI)?

Beyreuther

I am going to address this question in my talk. We clearly have patients who are very old, over 90, who have no Alzheimer pathology and who have no AAMI. And according to our studies, most of the AAMI is due to Alzheimer's or is associated or correlated with Alzheimer pathology and most of these cases are preclinical, too.

Gurling

The family with an amyloid mutation which was reported did have a case of onset with memory problems. I took the blood from the family initially and I remember they were very worried about the fact that one particular member could not operate the washing machine. That was the first symptom.

Mendlewicz

If I understand you correctly, all or most AAMI is early onset Alzheimer's, preclinical, presymptomatic Alzheimer's?

Beyreuther

You have to have those and if you do an age-matched control investigation, you start to screw up the system because you have a lot of preclinical, presymptomatic Alzheimer's.

Molecular Pathology and Aetiology of Alzheimer's Disease*

K. Beyreuther, C. Hilbich, G. König, G. Multhaup and C. L. Masters

Introduction

Alzheimer's disease (AD) is the most common cause of late-life intellectual impairment in countries that have achieved life expectancies above 70 years. It accounts for about 60% of the cases of dementia and its prevalence increases logarithmically with age. The clinical manifestations of AD include progressive impairments in short-term and — in later stages — in long-term memory, language, behaviour and disorientation in time and space (McKhann et al. 1986). At the cellular level, this dementia must be the result of neuronal dysfunction and degeneration. One can make then the assumption that this would lead to neuronal death, which is indeed a late event in AD (Terry et al. 1981; Mann et al. 1991). The AD brain shows a widespread but not uniform atrophy. Involved are mainly pyramidal neurons within the neocortical association area that provide cross connectivity of the association cortices. Markedly involved are also neurons in the entorhinal cortex, hippocampus, olfactory cortex, olfactory bulb, the basal nucleus of Meynert, the dorsal segmental nuclei, and the locus ceruleus nuclei (Rogers and Morrison 1985; Pearson et al. 1985). Relatively spared are brain stem, cerebellum, the motor cortex, visual and auditory cortices, and the greater part of the thalamus.

Abnormal proteinaceous deposits in the brain of AD patients meeting the ultrastructural as well as the histochemical criteria for amyloid define the disease at the molecular level. The application of a sensitive silver staining technique by Alzheimer described in his first report (Alzheimer 1907) revealed the presence of these abnormal structures within perikarya, axons, dendrites and terminals of neurons as neurofibrillary tangles (NFT) and in the extracellar neuropil as amyloid plaques (APC). In addition to the extracellular amyloid deposits of the APC, there is also a variable degree of extracellular amyloid around neocortical and meningeal blood vessels as amyloid congophilic angiopathy (ACA) in AD (Glenner and Wong 1984a,b). These fibrillar amyloid desposits are also found in the brains of all individuals with trisomy 21 (Down's syndrome, DS) after the age of 30 years (Rumble et al. 1989). The amyloid deposits tend to occur in the terminal zones of

*This work was supported by grants from the Deutsche Forschungsgemeinschaft through SFB 317 and 258, the Bundesminister für Forschung und Technologie of Germany, the Metropolitan Life Foundation, the Thyssen Stiftung, the Fonds der Chemischen Industrie of Germany, the Forschungsschwerpunkt BadenWürttemberg (to Konrad Beyreuther) and the National Health and Medical Research Council of Australia, the Victorian Health Promotion Foundation and the Aluminium Development Corporation of Australia (to Colin L. Masters).

neurons that undergo neurofibrillary degeneration and to spread along the neuronal projections of the pyramidal neurons within the association cortices, the hippocampus and parahippocampal structures (Hyman et al. 1984, 1986). Interestingly the amyloid deposition occurs bilaterally but not precisely symmetrically (Arendt et al. 1985), which is in agreement with interindividual variation in the clinical manifestation of AD.

We regard AD as a unitary phenomenon and a true disease, as opposed to "normal" aging in which clinical variables such as age at onset, rate of progression and severity of dementia are the result of the interaction of but a few critical steps in the generation of the AD amyloid subunit protein.

Since it has been shown that the occurrence of plaques, tangles and loss of synapses are linked with AD (Blessed et al. 1968; Davies et al. 1988; Hamos et al. 1989; Masliah et al. 1989; Rumble et al. 1989; DeKosky and Scheff 1990; Morris et al. 1991), the present challenge is to establish the role that amyloid fibril deposition plays in the process of AD and to address potential approaches to the postponement of its progression.

Amyloid Deposition and the Aetiology of Alzheimer's Disease

There are two extreme positions which define the ongoing controversy on the possible role that the proteinaceous amyloid deposition may play for the progressive neuronal and synaptic dysfunction and degeneration in AD. The first and still prominent position is that amyloid is a pathological by-product of the disease process and has no importance for the aetiology of AD. No matter how much we understand the genesis of the amyloid, it will never lead to any therapeutic insight in AD research. The second view is that the amyloid deposition is of fundamental significance in AD, and that a thorough understanding of the mechanism and pathogenesis of amyloid formation will eventually lead to a successful therapeutic intervention in AD.

Since we have been working on the nature of the amyloid in both AD and DS for the past 8 years, it is not difficult to guess which of these two extremes we favour. Before summarizing the arguments for and against these extreme positions, this review will first summarize the current state of knowledge on the nature, origin, genetics and the prevalence of the AD amyloid protein.

Amyloid ßA4 Protein as the Molecular Hallmark of Alzheimer's Disease

The principal component of the amyloid of AD is a protein that is now termed the ßA4 protein, according to its proposed secondary structure of ß-pleated sheets, and its relative molecular mass of 4 kDa (Beyreuther and Masters 1990, 1991; Hilbich et al. 1991a, b). Although some controversy still exists on the nature of the tangle, where not all observers agree with our interpretation of the evidence for the ßA4 protein in tangles (Masters et al. 1985a; Beyreuther et al. 1986; Beyreuther and Masters 1991) and argue that a microtubule-associated protein (tau) is the major, if not sole, component (Goedert et al. 1991; Mori and Ihara 1991), there is a broad unanimity that the ßA4 molecule is the major subunit of the extracellular amyloid deposits (Masters and Beyreuther 1991; Castano and Frangione 1991).

Protein sequencing of amyloid from brains of patients with AD as well as DS revealed the ßA4 protein to have a length of up to 42 to 43 residues (Glenner and Wong 1984a,b; Masters et al. 1985; Kang et al. 1987). The same ßA4 subunit is found in ACA (Glenner and Wong 1984a,b; Prelli et al. 1988), APC (Masters et al. 1985b), and NFT (Masters et al. 1985a; Beyreuther et al. 1986; Guiroy et al. 1987; Hyman et al. 1989). All kinds of amyloid ßA4 deposits contain considerable amounts of amino-terminal truncated peptides (Masters et al. 1985a,b; Beyreuther et al. 1986; Prelli et al. 1988). Carboxy-terminal truncated sequences were reported for ACA (Prelli et al. 1988). These findings were interpreted by us as indication that several proteases may be involved in the generation of ßA4 from its precursor (Masters et al. 1985a).

Electron microscopy has revealed a similar ultrastructure for APC and ACA. Besides minor amounts of other proteins (Rozemuller et al. 1989) and the non-specific neurites around some plaques, they contain ßA4 in the form of straight filaments having a diameter of 4-6 nm with a narrowing every 30-40 nm (Wisniewski 1983).

Amyloid Formation In Vitro

Synthetic ßA4 molecules adapt a ß-sheet conformation and form aggregates in vitro. These aggregates meet the ultrastructural as well as histochemical critera for amyloid (Hilbich et al. 1991a,b). The filaments formed by synthetic ßA4 have a diameter of 4.5-5.5 nm and built up dense networks that can be stained with the classic amyloid stain Congo red which subsequently show the green birefringence under polarized light. This tinctorial property is characteristic of naturally occurring amyloid and shows that amyloid and amyloid filaments similar to those found in the amyloid plaque of patients with AD can be assembled from chemically synthesized material under simple conditions in vitro. Filament formation of ßA4 proceeds without the need for any additional material such as aluminium ions, lipids or proteins, showing that the ßA4 sequence alone provides sufficient nucleation points for filament assembly. These results refute the widely discussed hypothesis that an additional nucleating component would be necessary to trigger the deposition of ßA4 filaments and would thereby represent a critical pathogenetic event in AD. Whilst other components of APC and ACA may facilitate the deposition of amyloid in AD, the major structural requirements for amyloid formation are contained within the ßA4 molecule (Hilbich et al. 1991a,b).

The biophysical basis for the aggregation of the ßA4 monomer into an amyloid fibril has been identified and resides in the hydrophobic interaction of critical regions of the ßA4 molecule (Hilbich et al. 1991a).

If amyloid deposition, as we believe, is of fundamental significance in AD, one can make the assumption that blocking the aggregation of ßA4 molecules in addition to approaches that aim to interfere with the process of de novo ßA4 generation would be an effective therapeutic strategy. The identification of residues important for the folding and aggregation of ßA4 by Hilbich et al. (1991a,b) provides the rationale for the proposed development of compounds that may inhibit ßA4 filament formation.

Site of Amyloid Deposition

Since ßA4 is highly insoluble and spontaneously self-assembles in vitro into amyloid fibrils, the release of ßA4 from its precursor is likely to occur close to the site of amyloid deposition. These sites are the neuropil, the perikarya of neurons and their neuritic processes, and less frequently the cerebral blood vessels. The formation of amorphous plaques is associated with a pathological change in the neuritic terminals. Presynaptic terminals associated with these amorphous ßA4 deposits are altered as shown by electron microscopic studies and immunostaining for the presynaptic marker synaptophysin (Brion et al. 1991; Bugiani et al. 1990; Yamaguchi et al. 1990). Synapses and neuronal membranes may be the initial sites for this process since these amorphous deposits accumulate around the soma of neurons and in close proximity of dendrites (Probst et al. 1991).

Amyloid ßA4 Protein in Incipient Alzheimer's Disease

It is clear that a true disease such as AD as opposed to normal aging must be associated with earliest pathological changes that occur before dementia can be reliably diagnosed clinically (Davies et al. 1988; Morris et al. 1991; Rumble et al. 1989). Therefore the presence of histopathological lesions characteristic for AD in nondemented cases may define patients with incipient AD and amyloid ßA4 protein may be used as a biological marker to identify patients at risk. A recent survey of the distribution and numbers of tangles and plaques in healthy elderly people and individuals with AD showed that as opposed to tangle formation amyloid ßA4 deposition is more directly indicative of a disease process with symptoms of dementia (J. L. Price et al. 1991). Tangles are found also in brains that do not contain deposits of amyloid ßA4 protein, and its formation is not specific for AD (Ghetti et al. 1989; J. L. Price et al. 1991). Since several studies reported that individuals with DS almost invariably develop the pathological changes associated with AD (Mann et al. 1989, 1991; Rumble et al. 1989), it has been argued that the pathoanatomical pattern of cases coming to autopsy before any cognitive changes have occurred would provide a description for the evolution of the amyloid pathology associated with AD. Autopsy studies revealed that the earliest change in DS involves a deposition of amyloid ßA4 protein within the association cortex (Rumble et al. 1989; Mann et al. 1991). Neuritic changes occur only later within some of the sites infiltrated with this amyloid (Rozemuller et al. 1989) and tangles may appear thereafter within parent nerve cell bodies (Hyman et al. 1984; Mann et al. 1989, 1991). Thus the evolution of the ßA4 amyloid lesion may start with deposition at extracellular sites in the brain which are normally occupied by nerve terminals and synapses and replace these structures.

Origin of Amyloid ßA4 Protein from the Transmembrane Amyloid Precursor Protein

The ßA4 amyloid arises from a much larger precursor protein, APP, also known as preA4 (Kang et al. 1987) (Fig. 1). This precursor of the amyloid ßA4 protein is encoded by the APP gene (also known as PAD gene) on chromosome 21 (Kang et al. 1987; Goldgaber et al. 1987; Salbaum et al. 1988).

The single copy APP (PAD) gene gives rise to a number of protein products. Alternative splicing leads to primary translation products of 770, 751, 714, 695 and 563 amino acids (DeSauvage and Octave 1989; Golde et al. 1990; Kang et al. 1987; Kang and Müller-Hill 1990; Kitaguchi et al. 1988; Ponte et al. 1988; Tanzi et al. 1988) (Fig. 1). The 563 amino acid product does not contain the ßA4 region which is encoded within exons 16 and 17 (Lemaire et al. 1989; Yoshikai et al. 1990). The latter product is designated as amyloid precursor rélated protein (APRP) in distinction to the amyloid precursor proteins (APP) which include the ßA4 sequence. The

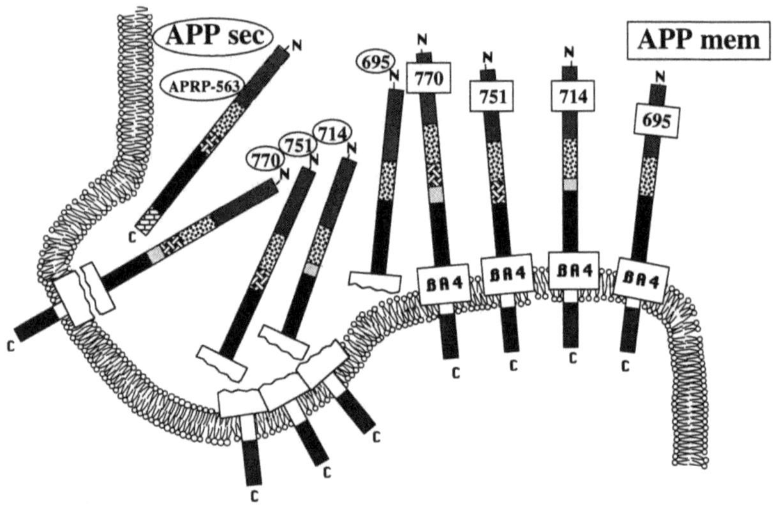

Fig. 1. Putative domain structure of transmembrane APP (*APP mem*) and its derivatives, the secretory APP forms (*APP sec*). Primary translational products are APP mem and the amyloid precursor related protein *APRP-563*. The intact ßA4 sequence is only included in the APP mem forms. The numbers refer to the length of the primary translational products (amino acid residues including the signal sequence). The extracellular domains are denoted by N (N-terminal part) and the cytoplasmic domain of APP mem by C (C-terminus). The proteolytic cleavage generating APP sec occurs within the ßA4 sequence C-terminal to the ßA4 residue 16. The domain corresponding to the Kunitz-type protease inhibitor (indicated by *crosses*) is present in APP_{751} and APP_{770}

other proteins are APPs and as such typical transmembrane proteins which span the lipid bilayer once (Dyrks et al. 1988; Weidemann et al. 1989). The larger forms (APP_{771}, APP_{751}) carry Kunitz-type serine protease inhibitory domains which in analogy to the protease inhibitor glia-derived neurite promoting factor (GDN; also termed protease nexin I, PN-1) may be involved in the regulation of the promotion of neurite outgrowth (Monard 1988). PN-1 activity depends on thrombin inhibition since thrombin brings about neurite retraction in cultured neuroblastoma cells.

Significance of Protease Inhibition by the Amyloid Precursor Protein

The secreted form of the Kunitz-type protease inhibitor-containing $APP_{751,770}$ which are produced by limited proteolysis from the corresponding transmembrane APP isoforms (Weidemann et al. 1989) is now known to be identical to a previously described molecule, protease nexin II (PN-2) (Oltersdorf et al. 1989; Van Nostrand et al. 1989) that inhibits blood clotting factor XIa , trypsin, chymotrypsin and the proteases termed epidermal growth factor binding protein and gamma subunit of nerve growth factor, NGF (Smith et al. 1990; Sinha et al. 1990).

Protease inhibitors and proteases have been proposed to be involved in many cellular functions, including receptor aggregation, neurite outgrowth, synaptic transmission and the maintenance of a long-lasting synaptic structure (Monard 1988; Rupp et al. 1991). The presence of APP forms with a protease inhibitor domain suggests therefore that these alternative splice products of nuclear APP pre mRNA may be involved in controlling the activity of degratory enzymes during development, regeneration and degeneration (Beyreuther and Masters 1991; König et al. 1991; Müller-Hill and Beyreuther 1989).

ßA4 Protein Is an Abnormal Proteolytic Breakdown Product

The ßA4 protein appears to be a proteolytic breakdown product of the APP precursor as predicted by us in 1985 (Masters et al. 1985a). Cleavage has to occur within the third extracellular domain that spans residues 200-624 of APP_{695} and within the transmembrane domain which corresponds to residues 625-648 of APP_{695} since the sequence of the ßA4 protein is identical with residues 596-638/639 of APP_{695} (Kang et al. 1987). The 42-43 amino acid ßA4 protein comprises half of the transmembrane domain encoded by exon 17 and the first 28 amino acids of the proximal extracellular domain encoded within exon 16 (Kang et al. 1987; Lemaire et al. 1989; Yoshikai 1990). Amyloid ßA4 protein is most likely not a primary translational product of an abnormally spliced APP mRNA form. Splicing of exon 1, which includes the translational start site to the ßA4 protein-encoding exons 16 and 17, would result in an mRNA which has several stop codons in its reading frame and which therefore could not encode the ßA4 protein (Lemaire et al. 1989).

Amyloid Precursor Proteins

The three most abundant APP transmembrane proteins ($APP_{695, 751, 770}$) are being studied extensively. In vitro translation studies monitored membrane insertion and the removal of the 17 amino acid signal peptide from the precursor during translation (Dyrks et al. 1988). The exact site of cleavage was determined by radiosequencing.

The precursor proteins undergo multiple post-translational modifications following signal peptide removal. We were able to demonstrate N-glycosylation, O-glycosylation, sulphation of tyrosine residues, and further proteolytic cleavage (Weidemann et al. 1989; Bunke et al. 1991). This cleavage dissects the extracellular domains of the APP from the transmembrane and cytoplasmic domains (Younkin 1991). The cleavage seems to occur at the cell surface and the shedded products are secretory APPs (APP_{sec} as opposed to APP_{mem}) that can be detected in conditioned media of cells expressing APP_{mem} (Weidemann et al. 1989). Recent evidence supports this hypothesis since the conversion to APP_{sec} can be inhibited by α2 macroglobulin added to cell cultures (Ganter et al. 1991). The half-life of APP_{mem} in cells is only 20-30 min and is one of the shortest reported so far (Weidemann et al. 1989).

The proteolytic cleavage of APP_{mem} in vitro in cultured cells occurs within the extracellular ßA4 domain (Figs. 1, 2). The precise site of the cleavage by APP secretase (Esch et al. 1990), more appropriately termed APPase, was determined to

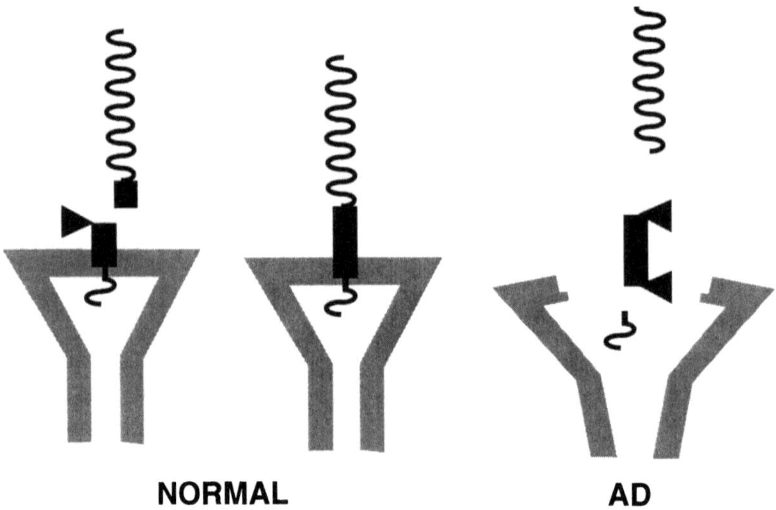

Fig. 2. Generation of amyloid ßA4 protein from the amyloid precursor protein (APP). The natural cleavage of APP by APPase occurs within the ßA4 sequence, thus neutralizing the amyloidogenic potential of transmembrane APP forms. APPase has a trypsin-like activity since it cleaves C-terminal of the ßA4 residue lysine 16. In AD the ßA4 amyloid subunit is produced by an alternative proteolytic process. This pathological processing of transmembrane APP requires two proteases whose specificity is different from that of APPase. Cleavage on the carboxyl side of the ßA4 occurs within the transmembrane domain and requires the release of this part from the membrane. The *solid bar* in each schematic indicates the position of the ßA4

be C-terminal to the ßA4-lysine residue 16 (APP_{695}-residue 611). Since this event occurs on the cell surface the two APP fragments produced are APP_{sec} and an 83 amino acid fragment that remains associated with the membrane (Esch et al. 1990; Sisodia et al. 1990; Weidemann et al. 1989; Younkin 1991). Thus this cleavage neutralizes the amyloidogenic potential of the amyloid precursors since both APP_{sec} and the 83 amino acid fragment are nonamyloidogenic proteins. APP_{sec} are found in serum and cerebrospinal fluid and other extracellular compartments (Bush et al. 1990; Mönning et al. 1990; Oltersdorf et al. 1989; Palmert et al. 1989; Rumble et al. 1989; Van Nostrand et al. 1989; Weidemann et al. 1989; Younkin 1991).

Since the precise site of cleavage to yield the normal secretory form of APP resides within the ßA4 protein segment of transmembrane APP, it is suggested that altered APPase activity may be a risk factor for AD and the APPase gene a candidate FAD (familial AD) locus. The genes of the proteases involved in the generation of the ßA4 protein are further candidate loci for FAD.

Normal Function of Amyloid Precursor Proteins

In the brain, where transmembrane APP is the major form (Schubert et al. 1991), the protein is subject to fast anterograde transport (Koo et al. 1990) and shown to undergo vesicular release at the level of the synapse (Schubert et al. 1991). The proposal that APP is a cell surface receptor (Kang et al. 1987) involved in the establishment and maintenance of cell-cell and cell-matrix interaction (Shivers et al. 1988) and thus crucial for brain plasticity is currently being explored.

The concept that APP binds to extracellular matrix (Klier et al. 1990; Narindrasorasak et al. 1991) is functionally appealing. If this is the case APP could act either as a transmembrane protein or as a secretory factor as signal for cell growth or targeting and attachment. The high concentration of APP in the alpha granule of the platelet (Bush et al. 1990; Van Nostrand et al. 1990) would suggest a function for APP in the coagulation cascade (Smith et al. 1990) and the wound healing and tissue damage repair process. Synthesis of APP by stimulated peripheral blood leucocytes (Bauer et al. 1991; Mönning et al. 1990) and reactive astrocytes (Siman et al. 1989) would also be consistent with a role in the repair process.

A general function and involvement of APP in regeneration and repair was assessed in the axotomy paradigm of degeneration and regeneration. APP expression changes in response to disconnection and reconnection of the neuron with its target (Scott et al. 1991). If regeneration is prevented, the post-axotomy response for APP expression persists in the affected neurons. This establishes that neuron-target interactions are important in APP gene regulation and is consistent with a role of APP in regeneration. The finding that neuron-target interactions are important in APP gene regulation together with the finding that the amyloidogenic transmembrane form of APP is a synaptic protein (Schubert et al. 1991; Shivers et al. 1988) must provide a clue to the restricted deposition of the ßA4 protein in brain.

Although there is probably only one function of APP in vivo there are many genes involved in assessing and controlling the normal APP function, which are all candidate genes and putative disease loci for AD. Among these are genes that are involved in regulation of APP expression, APP maturation, vesicular compartmentalization, transport and turnover of APP and genes encoding APP-binding proteins.

Regulation of the APP Gene

Overexpression of the APP gene may be involved in the pathogenesis of AD as suggested from the overproduction of the APP in DS (Rumble et al. 1989). We therefore analysed the structure of the putative regulatory region of APP (PAD) gene (Salbaum et al. 1988). The APP promoter has the characteristic features of promoters of housekeeping genes. It lacks a typical TATA box, has a high guanosine-cytosine (GC) content, and transcription initiates at multiple sites. Sites for DNA methylation are clustered in the region 400 bp upstream from the transcription start. The concept of a housekeeping gene is consistent with the ubiquitous expression of the APP gene. However, the term does not imply constitutive expression as the APP gene does respond to regulatory events (Mobley et al. 1988; König et al. 1990).

Following nerve injury macrophages invade the lesion site and secrete interleukin-1, which is known to be capable of increasing APP mRNA expression in vitro through acting at a responsive element at the APP promoter (Lindholm et al. 1987; Salbaum et al. 1988). Injury to nervous tissue elicits local inflammatory, glial, vascular and neuronal responses which have the potential to affect the expression of APP (Salbaum et al. 1988, 1990). After axotomy interleukin-1 can stimulate the local synthesis of NGF in non-neuronal cells at the site of nerve injury (Lindholm et al. 1987) and NGF is known to alter APP expression in vitro (König et al. 1990) and in vivo (Mobley et al. 1988). The expression of the heat shock protein HSP-70 is increased within 30 min after axotomy (New et al. 1989). The APP promoter includes a sequence homologous to the heat shock control element as well as two AP-1 consensus binding sites for the *fos/jun* heteroduplex (Salbaum et al. 1990). Increased expression for the two nuclear proteins c-*fos* and c-*jun* is also observed after nerve injury (Hengerer et al. 1990). Taken together the promoter structure includes elements that appear to be involved in regulating APP gene expression at sites of nerve injury. Since it has been shown that prevention of regeneration in the axotomy paradigm of degeneration and regeneration leads to a persistence of the post-axotomy response for APP expression (Scott et al. 1991) and since AD is characterized by disrupted neuron-target cell interaction (Hyman et al. 1986) and abortive regeneration (Probst et al. 1991) a local disturbance of APP gene regulation together with the production of superoxide by phagocytes could be the primary event leading to amyloid ßA4 protein formation.

It is also worth mentioning that perhaps aberrant regulation of the ratios of insert APP ($APP_{751/770}$) with protease inhibitor function to noninsert APP ($APP_{695/714}$) may play a role in the molecular aetiology of AD. The observed age-associated change in the splice ratios in favour of these insert APP forms (König et al. 1991) may be a risk factor for the development of amyloid deposits since the corresponding proteins are more resistant towards cleavage by APPase and thus increase the number of amyloidogenic APP_{mem} forms (Fig. 2). Therefore a decrease of insert-containing APP forms could potentially postpone the progression of amyloid deposition in AD. Retinoic acid has the properties of a factor for such a potential postponement strategy since we showed this compound to alter in vitro the splicing pattern in favour of the noninsert APP_{695} form (König et al. 1990).

Prevalence of ßA4 Protein Deposits and the Preclinical Period of Alzheimer's Disease

We have used immunocytochemical methods in a population-based survey of amyloid ßA4 protein deposition to estimate the time it takes to develop AD (Davies et al. 1988; Rumble et al. 1989). Our estimate of 30 years is in agreement with the age of the amyloid as judged from the rate of racemization of the amino acids aspartic acid and asparagine (Shapira et al. 1988).

Immunocytochemical studies of brain tissue from 26 patients with DS showed that the deposition of amyloid ßA4 protein began in these patients approximately 50 years earlier than it began in the normal-aging subjects (Rumble et al. 1989) (Figs. 3, 4). Deposits of ßA4 protein were detected in the cerebral cortex of younger Down's patients as early as 13 years of age, and by the end of the fifth decade in the brains of the normal-aging population (Davies et al. 1988). However, the rate of deposition for both groups was similar. If this rate of amyloid ßA4 protein deposition in DS, familial AD, sporadic AD and incipient AD (usually termed "normal aging") does not vary, one explanation may lie in the solubility of the ßA4 protein (Hilbich et al. 1991a,b). Assuming that a single initial event starts the pathological process by release of the ßA4 protein from the precursor, we would postulate that the spread of ßA4, which is suggested to occur along the projections of neurons, will simply depend on its solubility. Amyloid ßA4 protein is very poorly soluble in physiological buffers (Masters et al. 1985a,b; Hilbich et al. 1991a,b). This solubility will not be different in DS, familial AD and sporadic AD. We suggest this may be the main reason for the almost identical rate of ßA4 deposition in DS and AD.

The 30-year deposition and preclinical period in AD is consistent with the prevalence of conventionally diagnosed AD of 19% among persons aged 80-90 years and the prevalence of immunoreactive ßA4 protein deposits detected at necropsy in the brains of persons who are approximately 30 years younger (aged 50-60 years) (Fig. 3). In DS the prevalence of ßA4 protein deposition exceeds 50% among individuals aged 21-30 years and reaches 100% among individuals aged 30 years and older (Rumble et al. 1989) (Fig. 4). The incidence of clinically detectable dementia in DS was reported to exceed 50% among individuals aged 50-60 years and to be much higher among individuals with DS aged over 60 years (Lai and Williams 1989). This suggests that in DS the same 30 years of preclinical ßA4 protein accumulation precedes the onset of clinical symptoms of dementia.

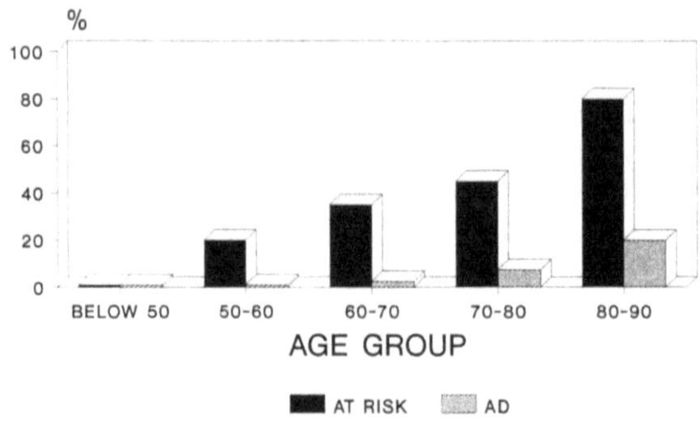

Fig. 3. Prevalence of cortical amyloid ßA4 deposits (*black columns*) and of post-mortem confirmed Alzheimer's disease (AD; *hatched columns*) in different age groups of normal karyotypes (years) (adapted from Davies et al. 1988 and Rumble et al. 1989). The presence of cortical ßA4 (at risk) includes patients with incipient and symptomatic AD

Mechanism of Amyloid Deposition in Alzheimer's Disease

We return to the question posed in the introduction to this paper of whether ßA4 amyloid deposition is fundamental in AD. Our current understanding of the mechanism of amyloid deposition is summarized in Figs. 2 and 5 where the action of an as yet unidentified protease generates the N-terminus of ßA4 as the initiating step. The membrane-associated ßA4 together with the cytoplasmic domain then undergoes a process of release from the membrane and second proteolytic cleavage to yield the variable C-terminus of ßA4 (Dyrks et al. 1988, 1991). At which stage the aggregation of ßA4 commences remains undetermined. The requirement of membrane damage for ßA4 amyloid generation links this process to neuronal and neuritic dysfunction and thus relates to the question of the significance of ßA4 deposition in AD.

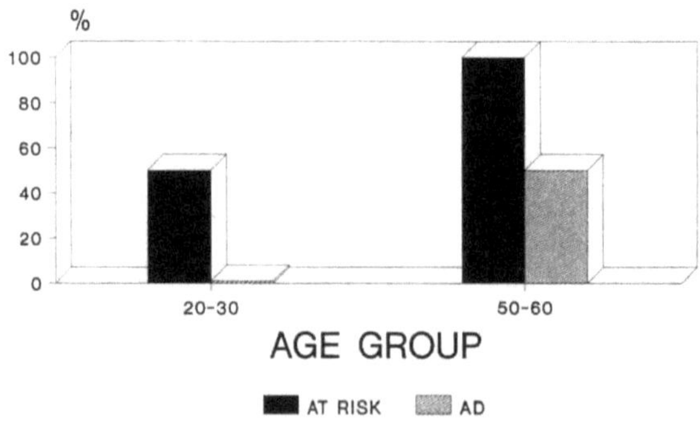

Fig. 4. Prevalence of cortical amyloid ßA4 deposits (*black columns*) and of post-mortem confirmed Alzheimer's disease (AD; *hatched columns*) in different age groups (years) of individuals with Down's syndrome (DS; adapted from Rumble et al. 1989). The mean age at onset for the process of amyloid ßA4 formation is 20-30 years in DS and 70-80 years in normal karyotypes (Fig. 3). For both groups 30 years later the same prevalence is found for the clinical diagnosis of AD which suggests that the duration of preclinical AD is approximately 30 years

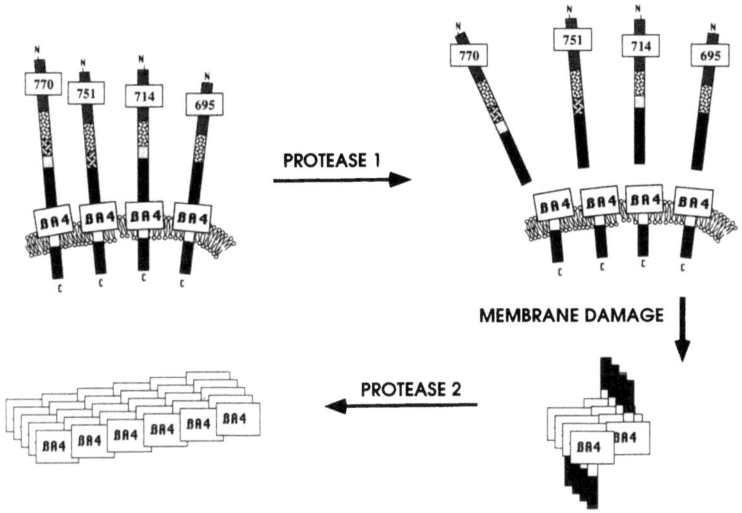

Fig. 5. Hypothetical model for the conversion of amyloid precursor proteins (APP) to ßA4 amyloid in human brain. According to this model a protease cleaves APP on the amino site of ßA4 to generate the fragment A4CT (Dyrks et al. 1988). Only after disruption of the membrane can this fragment be released to allow aggregation. A second protease then releases the non-ßA4 segment of A4CT to leave the highly insoluble and fibril-forming ßA4 amyloid subunit.

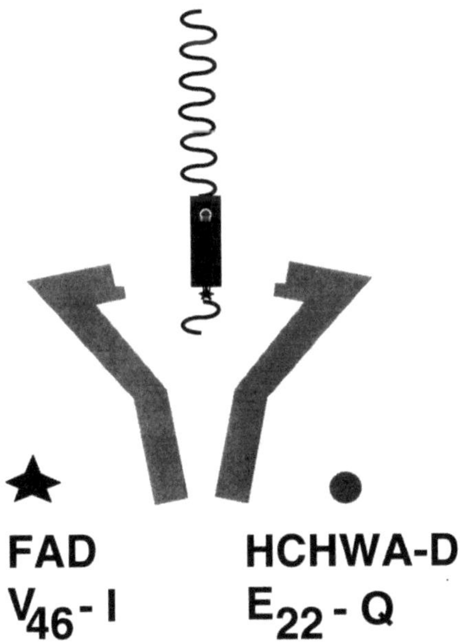

Fig. 6. Disease-associated mutations have been identified in the locus linked to some forms of familial Alzheimer's disease (*FAD*) and to hereditary cerebral hemorrhage with amyloidosis (Dutch) (*HCHWA-D*). The mutation in a Dutch familiy with congophilic angiopathy changes the glutamic acid residue 22 (*E*) of the ßA4 region into glutamine (*Q*). In several families with AD a mutation at position 46 of the ßA4 region which changes the natural valine residue (*V*) into isocleucine (*I*) has been linked with the disease. The *solid bar* indicates the position of the ßA4.

The report of a linkage between chromosome 21 and FAD has led to a productive search for mutations in the APP locus linked to FAD (St. George-Hyslop et al. 1987). In a Dutch family with congophilic angiopathy (HCHWA-D), a disease-associated mutation at position 22 of the ßA4 region (Fig. 6) has been identified (Levy et al. 1990; Van Broeckhoven et al. 1990), and in an increasing number of families with AD a mutation at position 46 of the ßA4 region (Fig. 6) ($APP_{695/751/770}$-residue 542, 598 and 717, respectively) has been linked with the disease (Goate et al. 1991; Naruse et al. 1991; Yoshioka et al. 1991). These mutations provide compelling evidence for the central role that APP and its pathologic breakdown product ßA4 must play in the pathogenesis of AD. Our task is now to understand how these mutations affect the processing of APP into ßA4. We also anticipate that other mutations linked to FAD which may map outside the APP gene will affect the processing of APP into the amyloid subunit.

Although not yet defined, we suspect that the synaptic and vesicular localization of APP and the elucidation of the normal function of APP in the brain will prove to be critical determinants in understanding the process by which neurons degenerate in AD.

The future search for factors that induce and those that inhibit the formation of ßA4 depends on adequate in vitro and in vivo models. Promising starts have been made with synthetic amyloid ßA4 peptides (Hilbich et al. 1991a,b), expression of APP fragments of the A4CT type (Dyrks et al. 1988, 1991; Maruyama et al. 1990) (Fig. 6), and APP or APP fragments overexpressing transgenic mice which are currently being investigated in our laboratories (Beer et al. 1991). The formation of amyloid ßA4 protein deposits in brains of transgenic mice has been reported and may greatly facilitate the elucidation of the mechanism of amyloid formation and its role in the neurodegenerative processes of AD (Quon et al. 1991; Wirak et al. 1991). Another very promising in vivo model is the simulation of DS in mice with the manipulation of trisomic 16 mouse brain tissues yielding changes which mimic those of AD (Richards et al. 1991). These models will complement the existing animal models of aged nonhuman primates and dogs (D. L. Price et al. 1991).

The real test of our hypothesis of a fundamental significance of ßA4 for AD will come when therapeutic strategies are developed around the concept of preventing the generation of ßA4 amyloid or of removing the earliest para-amyloid aggregates (Davies et al. 1988; Yamaguchi et al. 1988) from the brain. We estimate that the preclinical phase of AD could extend over a period of 30 or more years (Davies et al. 1988; Rozemuller et al. 1989; Rumble et al. 1989), which could present a formidable problem in applying any therapeutic strategy directed at these structural lesions.

References

Alzheimer A (1907) Über eine eigenartige Erkrankung der Hirnrinde. Zentralbl Nervenheilk Psychiatr 18: 177-179

Arendt T, Bigl V, Tennstedt A, Arendt A (1985) Neuronal loss in different parts of the nucleus basalis is related to neuritic plaque formation in cortical target areas in Alzheimer's disease. Neuroscience 14: 1-9

Bauer J, König G, Strauss S, Jonas U, Ganter U, Weidemann A, Mönning U, Masters CL, Volk B, Berger M, Beyreuther K (1991) In-vitro matured human macrophages express Alzheimer's ßA4-amyloid precursor protein indicating synthesis in microglial cells. FEBS Lett 282:335-340

Beer J, Salbaum JM, Schlichtmann E, Hoppe P, Earley S, Carlson GA, Masters CL, Beyreuther K (1991) Transgenic mice and Alzheimer's disease. In: Iqbal K, McLachlan DRC, Winblad B, Wisniewski HM (eds) Alzheimer's disease: basic mechanisms, diagnosis and therapeutic strategies. Wiley, Chichester, pp 473-478

Beyreuther K, Masters CL (1990) Nomenclature of amyloid A4 proteins and their precursors in Alzheimer's disease and Down's syndrome. Neurobiol Aging 11:66-68

Beyreuther K, Masters CL (1991) Amyloid precursor protein (APP) and ßA4 amyloid in the etiology of Alzheimer's disease: precursor-product relationships in the derangement of neuronal function. Brain Pathol 1:241-251

Beyreuther K, Multhaup G, Simms G, Pottgiesser J, Schröder W, Martins RN, Masters CL (1986) Neurofibrillary tangles of Alzheimer disease and "aged" Down's syndrome contain the same protein as the amyloid of plaque cores and blood vessels. Discuss Neurosci 3:68-80

Blessed G, Tomlinson E, Roth M (1968) The association between quantitative measures of dementia and of senile change in the cerebral grey matter of elderly subjects. Br J Psychiatry 114:797-811s

Brion JP, Couk AM, Bruce M, Anderton B, Flamont-Durand J (1991) Synaptophysin and chromogranin A immunoreactivity in senile plaques of Alzheimer's disease. Brain Res 539:143-150

Bugiani O, Giacone G, Verga L, Pollo B, Ghetti B, Frangione B, Tagliavini F (1990) Alzheimer patients and Down patients: abnormal presynaptic terminals are related to cerebral pre-amyloid deposits. Neurosci Lett 119:56-59

Bunke D, Mönning U, Kypta RM, Courtneidge SA, Masters CL, Beyreuther K (1991) Tyrosine phosphorylation of the cytoplasmic domain of Alzheimer amyloid protein precursors. In: Iqbal K, McLachlan KDRC, Winblad B, Wisniewski HM (eds) Alzheimer's disease: basic mechanisms, diagnosis and therapeutic strategies. Wiley, New York, pp 229-235

Bush AI, Martins RN, Rumble B, Moir R, Fullser S, Milward E, Currie J, Ames D, Weidemann A, Fischer P, Multhaup G, Beyreuther K, Masters CL (1990) The amyloid precursor protein of Alzheimer's disease is released by human platelets. J Biol Chem 265:15977-15983

Castano EM, Frangione B (1991) Alzheimer's disease from the perspective of the systemic and localized forms of amyloidosis. Brain Pathol 1:263-271

Davies L, Wolska B, Hilbich C, Multhaup G, Martins R, Simms G, Beyreuther K, Masters CL (1988) A4 amyloid protein deposition and the diagnosis of Alzheimer's disease: prevalence in aged brains determined by immunocytochemistry compared with conventional neuropathologic techniques. Neurology 38:1688-1693

DeKosky ST, Scheff SW (1990) Synapse loss in frontal cortex biopsies in Alzheimer's disease: correlation with cognitive severity. Ann Neurol 27:457-464

DeSauvage F, Octave J-N (1989) A novel mRNA of the A4 amyloid precursor gene coding for a possibly secreted protein. Science 245:651-653

Dyrks T, Weidemann A, Multhaup G, Salbaum JM, Lemaire HG, Kang J, Müller-Hill B, Masters CL, Beyreuther K (1988) Identification, transmembrane orientation and biogenesis of the amyloid A4 precursor of Alzheimer's disease. EMBO J 7:949-957

Dyrks T, Mack E, Masters CL, Beyreuther K (1991) Membrane insertion prevents aggregation of precursor fragments containing the ßA4 sequence of Alzheimer's disease. In: Iqbal K, McLachlan DRC, Winblad B, Wisniewski HM (eds) Alzheimer's disease: basic mechanisms, diagnosis and therapeutic strategies. Wiley, Chichester, pp 281-287

Esch GS, Keim PS, Beattie EC, Blacher RW, Culwell AR, Oltersdorf T, McClure D, Ward PJ (1990) Cleavage of amyloid ß peptide during constitutive processing of its precursor. Science 248:1122-1124

Ganter U, Strauss S, Jonas U, Weidemann A, Beyreuther K, Volk B, Berger M, Bauer J (1991) Alpha 2-macroglobulin synthesis in interleukin-6-stimulated human neuronal (SH-SY5Y neuroblastoma) cells: potential significance for the processing of Alzheimer ß-amyloid precursor protein. FEBS Lett 282:127-131

Ghetti B, Tagliavini F, Masters CL, Beyreuther K, Giaccone G, Vega L, Farlow MR, Conneally PM, Dlouhy SR, Azzarelli B, Bugiani O (1989) Gerstmann-Sträussler-Scheinker disease. II. Neurofibrillary tangles and plaques with PrP-amyloid coexist in an affected family. Neurology 39:1453-1461

Glenner GG, Wong CW (1984a) Alzheimer's disease: initial report of the purification and characterization of a novel cerebrovascular amyloid protein. Biochem Biophys Res Commun 120:885-890
Glenner GG, Wong CW (1984b) Alzheimer's disease and Down's syndrome sharing of a unique cerebrovascular amyloid fibril protein. Biochem Biophys Res Commun 122:1131-1135
Goate A, Chartier-Harlin M-C, Mullan M, Brown J, Crawford F, Fidani L, Giuffra L, Haynes A, Irving N, James L, Mant R, Newton P, Rooke K, Roques P, Talbot C, Pericak-Vance M, Roses A, Williamson R, Rossor M, Owen M, Hardy J (1991) Segregation of a missense mutation in the amyloid precursor protein gene with familial Alzheimer's disease. Nature 349:704-706
Goedert M, Spillantini MG, Crowther RA (1991) Tau proteins and neurofibrillary degeneration. Brain Pathol 1:279-286
Golde TE, Estus S, Usiak M, Younkin LH, Younkin SG (1990) Expression of amyloid protein precursor mRNAs: recognition of a novel alternatively spliced form and quantitation in Alzheimer's disease using PCR. Neuron 4:253-267
Goldgaber D, Lerman MI, McBride OW, Saffiotti U, Gajdusek DC (1987) Characterization and chromosomal localization of a cDNA encoding brain amyloid of Alzheimer's disease. Science 235:877-880
Guiroy DC, Miyazaki M, Multhaup G, Fischer P, Garruto RM, Beyreuther K, Masters CL, Simms G, Gibbs CJ Jr, Gajdusek DC (1987) Amyloid of neurofibrillary tangles of Guamanian parkinsonism-dementia and Alzheimer disease share identical amino acid sequence. Proc Natl Acad Sci USA 84:2073-2077
Hamos JE, DeGennaro LJ, Drachman DA (1989) Synaptic loss in Alzheimer's disease. Neurology 39:355-361
Hengerer B, Lindholm D, Heumann R, Rüther U, Wagner F, Thoenen H (1990) Lesion induced increase in nerve growth factor mRNA is mediated by c-fos. Proc Natl Acad Sci USA 87:3899-3903
Hilbich C, Kisters-Woike B, Reed J, Masters CL, Beyreuther K (1991a) Aggregation and secondary structure of synthetic amyloid ßA4 peptides of Alzheimer's disease. J Mol Biol 218:149-163
Hilbich C, Kisters-Woike B, Reed J, Masters CL, Beyreuther K (1991b) Human and rodent sequence analogs of Alzheimer's amyloid ßA4 share similar properties and can be solubilized in buffers of pH 7.4. Eur J Biochem 201:61-69
Hyman BT, Van Hoesen GW, Damasio AR, Barnes CL (1984) Alzheimer's disease: cell specific pathology isolates the hippocampal formation. Science 225:1168-1170
Hyman BT, Van Hoesen GW, Kromer LJ, Damasio AR (1986) Perforant pathway changes and the memory impairment of Alzheimer's disease. Ann Neurol 20:472-481
Hyman BT, Van Hoesen GW, Beyreuther K, Masters CL (1989) A4 amyloid protein immunoreactivity is present in Alzheimer's disease neurofibrillary tangles. Neurosci Lett 110:352-355
Kang J, Müller-Hill B (1990) Differential splicing of Alzheimer's disease amyloid A4 precursor RNA in rat tissues: PreA4695 is predominantly produced in rat and human brain. Biochem Biophys Res Commun 166:1192-1200
Kang J, Lemaire H-G, Unterbeck A, Salbaum JM, Masters CL, Grzeschik K-H, Multhaup G, Beyreuther K, Müller-Hill B (1987) The precursor of Alzheimer's disease amyloid A4 protein resembles a cell surface receptor. Nature 325:733-736
Kitaguchi N, Takahashi Y, Tokushima Y, Shiojiri S, Ito H (1988) Novel precursor of Alzheimer's disease amyloid protein shows protease inhibitory activity. Nature 331:530-532
Klier FG, Cole G, Stallcup W, Schubert D (1990) Amyloid ß-protein precursor is associated with extracellular matrix. Brain Res 515:336-342
König G, Masters CL, Beyreuther K (1990) Retinoic acid induced differentiated neuroblastoma cells show increased expression of the ßA4 amyloid gene of Alzheimer's disease and an altered splicing pattern. FEBS Lett 269:305-310
König G, Salbaum JM, Wiestler O, Lang W, Schmitt HP, Masters CL, Beyreuther K (1991) Alternative splicing of the ßA4 amyloid gene of Alzheimer's disease in cortex of control and Alzheimer's disease patients. Mol Brain Res 9:259-262
Koo EH, Sisodia SS, Archer DR, Martin LJ, Weidemann A, Beyreuther K, Fischer P, Masters CL, Price DL (1990) Precursor of amyloid protein in Alzheimer disease undergoes fast anterograde axonal transport. Proc Natl Acad Sci USA 87:1561-1565
Lai F, Williams RS (1989) A prospective study of Alzheimer's disease in Down syndrome. Arch Neurol 46:849-853
Levy E, Carman MD, Fernandez-Madrid IJ, Power MD, Lieberburg I, Van Duinen SG, Bots GTA, Luyendijk W, Frangione B (1990) Mutation of the Alzheimer's disease amyloid gene in hereditary cerebral hemorrhage, Dutch type. Science 248: 1124-1126

Lemaire H-G, Salbaum JM, Multhaup G, Kang J, Bayney RM, Unterbeck AK, Beyreuther K, Müller-Hill B (1989) The PreA4$_{695}$ precursor protein of Alzheimer's disease A4 amyloid is encoded by 16 exons. Nucleic Acids Res 17:517-522

Lindholm D, Heumann R, Meyer M, Thoenen H (1987) Interleukin-1 regulates synthesis of nerve growth factor in non-neuronal cells or rat sciatic nerve. Nature 330:658-659

Mann DMA, Brown A, Prinja D, Davies CA, Landon M, Masters CL, Beyreuther K (1989) An analysis of the morphology of senile plaques in Down's syndrome patients of different ages using immunocytochemical and histochemical techniques. Neuropathol Appl Neurobiol 15:317-329

Mann DMA, Jones D, Prinja D, Purkiss M (1991) Amyloid deposits within the cerebrum and cerebellum in Alzheimer's disease and Down's syndrome. Acta Neuropathol (Berl) 80:318-327

Marayuma K, Terakado K, Usami M, Yoshikawa K (1990) Formation of amyloid-like fibrils in COS cells overexpressing part of the Alzheimer amyloid protein precursor. Nature 347:566-569

Masliah E, Terry RD, DeTeresa RM, Hansen LA (1989) Immunohistochemical quantification of the synapse-related protein synaptophysin in Alzheimer's disease. Neurosci Lett 103:234-239

Masters CL, Beyreuther K (1991) Alzheimer's disease: molecular basis of structural lesions. Brain Pathol 1:226-227

Masters CL, Multhaup G, Simms G, Pottgiesser J, Martins RN, Beyreuther K (1985a) Neuronal origin of a cerebral amyloid: neurofibrillary tangles of Alzheimer's disease contain the same protein as the amyloid of plaque cores and blood vessels. EMBO J 4:2757-2763

Masters CL, Simms G, Weinman NA, Multhaup G, McDonald BL, Beyreuther K (1985b) Amyloid plaque core protein in Alzheimer's disease and Down syndrome. Proc Natl Acad Sci USA 82:4245-4249

McKhann G, Drachmann D, Folstein M, Katzman R, Price D, Stadlan EM (1986) Clinical diagnosis of Alzheimer's disease: report of the NINCDS-ADRA Work Group under the auspices of Department of Health and Human Services Task Force on Alzheimer's disease. Neurology 34:939-944

Mobley WC, Neve RL, Prusiner SB, McKinley MP (1988) Nerve growth factor increases mRNA levels for the prion protein and the ß-amyloid protein. Proc Natl Acad Sci USA 87:2405-2408

Monard D (1988) Cell-derived proteases and protease inhibitors as regulators of neurite outgrowth. Trends Neurosci 11:541-544

Mönning U, König G, Prior R, Mechler H, Schreiter-Gasser U, Masters CL, Beyreuther K (1990) Synthesis and secretion of Alzheimer amyloid ßA4 precursor protein by stimulated human peripheral blood leucocytes. FEBS Lett 277:261-266

Mori H, Ihara Y (1991) Neurofibrillary tangles, dystrophic neurites (curly fibers), and abnormal phosphorylation of tau. Brain Pathol 1:273-277

Morris JC, McKeel DW, Storandt M, Rubin EH, Price JL, Grant EA, Ball MJ, Berg L (1991) Very mild Alzheimer's disease: informant-based clinical, psychometric, and pathologic distinction from normal aging. Neurology 41:469-478

Müller-Hill B, Beyreuther K (1989) Molecular biology of Alzheimer's disease. Annu Rev Biochem 58:287-307

Narindrasorasak S, Lowery D, Gonzales-DeWhitt P, Poorman RA, Greenberg B, Kisilevsky R (1991) High affinity interactions between the Alzheimer's ß-amyloid precursor proteins and the basement membrane forms of heparan sulfate proteoglycan. J Biol Chem 266:12878-12883

Naruse S, Igarashi S, Kobayashi H, Aoki K, Inuzuka T, Kaneko K, Shimizu T, Iihara K, Kojinia T, Miyatake T, Tsuji S (1991) Mis-sense mutation val ---> Ile in exon 17 of amyloid precursor protein gene in Japanese familial Alzheimer's disease. Lancet 337:978-979

New RL, Hendricksen BR, Jones KJ (1989) Induction of heat shock protein 70 mRNA in adult hamster facial nuclear groups following axtomoy of the facial nerve. Metabol Brain Dis 4:273-279

Oltersdorf T, Fritz LC, Schenk DB, Lieberburg I, Johnson-Wood KL, Beatti EC, Ward PJ, Blacher RW, Dovey HF, Sinha S (1989) The secreted form of the Alzheimer's amyloid precursor protein with the Kunitz domain is protease nexin-II. Nature 341:144-147

Palmert MR, Podlisny MB, Witker DS, Oltersdorf T, Younkin LH, Selkoe DJ, Younkin SG (1989) The ß-amyloid protein precursor of Alzheimer disease has soluble derivatives found in human brain and cerebrospinal fluid. Proc Natl Acad Sci USA 86:6338-6342

Pearson RCA, Esiri MM, Hiorns RW, Wilcock GK, Powell TPS (1985) Anatomical correlates of the distribution of the pathological changes in the neocortex in Alzheimer's disease. Proc Natl Acad Sci USA 82:4531-4534

Ponte P, Gonzalez-DeWhitt P, Schilling J, Miller J, Hsu D, Greenberg B, Davis K, Wallace W, Lieberburg I, Fuller F, Cordell B (1988) A new A4 amyloid mRNA contains a domain homologous to serine proteinase inhibitors. Nature 311:525-527

Prelli F, Castano E, Glenner GG, Frangione B (1988) Differences between vascular and plaque core amyloid in Alzheimer's disease. J Neurochem 51:648-651

Price JL, Davis PB, Morris JC, White DL (1991a) The distribution of tangles, plaques and related immunohistochemical markers in healthy aging and Alzheimer's disease. Neurobiol Aging 12:295-312

Price DL, Martin LJ, Sisodia SS, Wagster MV, Koo EH, Walker LC, Koliatsos VE, Cork LC (1991b) Aged non-human primates: an animal model of age-associated neuro-degenerative disease. Brain Pathol 4: 287-296

Probst A, Langui D, Ulrich J (1991) Alzheimer's disease: a desrciption of the structural lesions. Brain Pathol 1:229-239

Quon D, Wang Y, Catalano R, Scardina JM, Murakami K, Cordell B (1991) Formation of ß-amyloid protein deposits in brains of transgenic mice. Nature 352:239-241

Richards S-J, Waters JJ, Wischik CM, Abraham CR, Sparkman DR, White CL, Beyreuther K, Masters CL, Dunnett SB (1991) Transplants of mouse trisomy 16 hippocampus provide an in vivo model of the neuropathology of Alzheimer's disease. EMBO J 7:297-303

Rogers J, Morrison JH (1985) Quantitative morphology and regional and laminar distributions of senile plaques in Alzheimer's disease. J Neurosci 5:2801-2808

Rozemuller JM, Eikelenboom P, Stam FC, Beyreuther K, Masters CL (1989) A4 Protein in Alzheimer's disease: primary and secondary cellular events in extracellular amyloid deposition. J Neuropath Expl Neurol 48:674-691

Rumble B, Retallak R, Hilbich C, Simms G, Multhaup G, Martins R, Hockey A, Montgomery P, Beyreuther K, Masters CL (1989) Amyloid A4 protein and its precursor in Down's syndrome and Alzheimer's disease. N Engl J Med 320:1446-1452

Rupp F, Payan DG, Magill-Solc C, Cowan DM, Scheller RH (1991) Structure and expression of a rat agrin. Neuron 6:811-823

Salbaum JM, Weidemann A, Lemaire H-G, Masters CL, Beyreuther K (1988) The promoter of Alzheimer's disease amyloid A4 precursor gene. EMBO J 7:2807-2813

Salbaum JM, König G, Beer J, Multhaup G, Masters CL, Beyreuther K (1990) Regulation of the amyloid gene of Alzheimer's disease. In: Beyreuther K, Schettler G (eds) Molecular biology of aging. Springer, Berlin Heidelberg New York, pp 89-96

Schubert W, Prior R, Weidemann A, Dircksen H, Multhaup G, Master CL, Beyreuther K (1991) Localization of Alzheimer ßA4 amyloid precursor protein at central and peripheral synaptic sites. Brain Res (in press)

Scott JN, Parhard I, Clark AW (1991) ß-Amyloid precursor protein gene is differentially epxressed in axotomized sensory and motor systems. Mol Brain Res 10:315-325

Shapira R, Austin GE, Mira SS (1988) Neuritic plaque amyloid in Alzheimer's disease is highly racemized. J Neurochem 50:69-74

Shivers BD, Hilbich C, Multhaup G, Salbaum M, Beyreuther K, Seeburg PH (1988) Alzheimer's disease amyloidogenic glycoprotein: expression pattern in rat brain suggests role in cell contact. EMBO J 7: 1365-1370

Siman R, Card JP, Nelson RB, Davis LG (1989) Expression of ß amyloid precursor protein in reactive astrocytes following neuronal damage. Neuron 3:275-285

Sinha S, Dovey HF, Seubert P, Ward PJ, Blacher RW, Blaber M, Bradshaw RA, Arici M, Mobley WC, Lieberburg I (1990) The protease inhibitory properties of the Alzheimer's ß-amyloid precursor protein. J Biol Chem 265:8983-8955

Sisodia SS, Koo EH, Beyreuther K, Unterbeck A, Price DL (1990) Evidence that ß-amyloid protein in Alzheimer's disease is not derived by normal processing. Science 248:492-495

Smith RP, Higuchi DA, Broze GJ Jr (1990) Platelet coagulation factor XI_a-inhibitor, a form of Alzheimer amyloid precursor protein. Science 248:1126-1128

St. George-Hyslop PH, Tanzi RE, Polinsky RJ, Haines JL, Nee L, Watkins PC, Myers RH, Feldman RG, Pollen D, Drachman D, Growdon J, Bruni A, Foncin J-F, Salmon D, Frommelt P, Amaducci L, Sorbi S, Piacenti S, Stewart GD, Hobbs WJ, Conneally PM, Gusella JF (1987) The genetic defect causing familial Alzheimer's disease maps on chromosome 21. Science 235:885-890

Tanzi RE, McClatchey AI, Lamperti ED, Villa-Komaroff L, Gusella JF, Neve RL (1988) Protease inhibitor domain encoded by an amyloid protein precursor mRNA associated with Alzheimer's disease. Nature 311:528-530

Terry RD, Peck A, DeTeresa R, Schechter R, Horoupian DS (1981) Some morphometric aspects of the brain in senile dementia of the Alzheimer type. Ann Neurol 10:184-192

Van Broeckhoven C, Haan J, Bakker E, Hardy JA, Van Hul W, Wehnert A, Vegter-Van der Vlis M, Roos RAC (1990) Amyloid ß protein precursor gene and hereditary cerebral hemorrhage with amyloidosis (Dutch). Science 248:1120-1122

Van Nostrand WE, Wagner SL, Suzuki M, Choi BH, Farrow JS, Geddes JW, Cotman CW, Cunningham DD (1989) Protease nexin-II, a potent antichymotrypsin, shows identity to amyloid ß-protein precursor. Nature 341:546-549

Van Nostrand WE, Schaier AH, Farrow JS, Cunningham DD (1990) Protease nexin-II (amyloid ß-protein precursor): a platelet alpha-granule protein. Science 248:745-748

Weidemann A, König G, Bunke D, Fischer P, Salbaum JM, Masters CL, Beyreuther K (1989) Identification, biogenesis and localization of precursors of Alzheimer's disease A4 amyloid protein. Cell 57:115-126

Wirak DO, Byney R, Ramabhadran TV, Fracasso RP, Hart JT, Hauer PE, Hsiau P, Pekar SK, Scangos BD, Unterbeck AJ (1991) Deposits of amyloid ß protein in the central nervous system of transgenic mice.

Wisniewski HM (1983) Neuritic (senile) and amyloid plaques. In: Reisberg B (ed) Alzheimer's disease. The Free Press, New York, pp 57-61

Yamaguchi H, Hirai S, Morimatsu M, Shoji M, Ihara Y (1988) A variety of cerebral amyloid deposits in the brains of the Alzheimer-type dementia demonstrated by ß protein immunostaining. Acta Neuropathol 76:541-549

Yamagushi H, Nakazato Y, Hirai S, Shoji M (1990) Immuno-electron microscopic localization of amyloid ß protein in the diffuse plaques of the Alzheimer-type dementia. Brain Res 508:593-597

Yoshikai S, Sasaki H, Doh-ura K, Furuya H, Sasaki Y (1990) Genomic organization of the human amyloid beta-protein precursor gene. Gene 87:257-263

Yoshioka K, Miki T, Katsuya T, Ogihara T, Sasaki Y (1991) The ^{717}Val-Ile substitution in amyloid precursor protein is associated with familial Alzheimer's disease regardless of ethnic groups. Biochem Biophys Res Comm 178:1141-1146

Younkin SG (1991) Processing of the Alzheimer's disease ßA4 amyloid protein precursor (APP). Brain Pathol 1:253-262

Discussion of Presentation K. Beyreuther et al.

Tsuang
Selkoe's group in Boston reported about skin fibroblast amyloid deposition. What is the current status of their finding and to what extent could this be utilized for the diagnosis of preclinical Alzheimer's disease?

Beyreuther
This was a very interesting finding. In Down's syndrome and some Alzheimer patients, but not in controls, Selkoe's group found amyloid deposits in skin and rectal biopsy tissue, which more or less means fibroblasts. Everybody has tried to reproduce this, even at his own company, Athena neuroscience. However, nobody, including our group, was able to reproduce this result. But we all may not be good enough in handling the corresponding antibody. This is very difficult. The problem is that the amyloid antibodies that we all used react with the amyloid precursor. Since the amyloid precursor is made by almost all cells, we see immunoreactivity everywhere, but this is not specific for amyloid deposition. If you try to identify the antigen, the amyloid is not found, only the amyloid precursor. So I would say, there is a problem with the technique. Alzheimer's is a disease and as such not normal aging. It has a clear pathology. You expect to have a pathology before you see symptoms. So you would expect to have a lot of people who are old who show that skin reactivity without having Alzheimer's disease. Selkoe's group claims to detect skin amyloid immunoreactivity only in patients who have the disease, i.e., clinical symptoms, so something is wrong. There is no disease which does not have a preclinical stage, and there is no biological disease marker that you do not see at preclinical stages.

Reich
Do you have any direct evidence of studies of the brains of non-Alzheimer siblings or other high-risk individuals to compare to normals who do not become affected?

Beyreuther
Among the people who died between age 70 and 80 years about 25%–30% had no Alzheimer pathology. There are people of old age who do not show Alzheimer amyloid pathology. Unfortunately we do not have twin data available right now. I do not know of any study that addresses that question. This also holds true for relatives of individuals with preclinical conditions and people at risk but unaffected dying from other causes.

Van Broeckhoven
I do not know of any results, but you should remember that it is very difficult to persuade the family of a nonaffected individual or a sibling of an affected parent to have postmortem examination. It is difficult enough to persuade families to have a postmortem examination of a patient.

Reich
My understanding of the work in St. Louis is that you can obtain brain from spouses and other related individuals as controls and I think it is a matter of effort rather than possibility to get the brains from siblings who may be dying of cancer or any other disorder in order to do a high-risk study. It seems to me that you want to quantitate a phenotype, an underlying phenotype, and that phenotype cannot be studied from such material.

Beyreuther
We have started to establish such a cohort together in collaboration with Dr. Häfner's hospital (ZISG) in Mannheim. By looking at possible cases of familial Alzheimer's disease we found that one of the most promising families had mixed pathology, vascular and Alzheimer pathology. It is very difficult to find good families. We have collected more families which we are planning to study but this may take years. Furthermore controls are very difficult to get.

Vogel
Of course the situation in United States, i.e., the psychological situation of the families, is quite different from that in Germany. It is much more difficult in Europe, not only in Germany, to convince people of postmortem investigations than it is in the US. But another question, you did not say anything about the molecular bases of those dominantly inherited types of Alzheimer's disease in which this gene on chromosome 21 is unaffected, and there are many such families. Can you say anything about the molecular basis of these families?

Beyreuther
I can only comment on pathology. We have looked at many of those cases and we do not see any difference with regard to pathology. Even in Down's syndrome, people always make the distinction that there may be a different progression of the amyloid pathology in Down's syndrome and normal karyotypes. But the progression we found in our cohort study was the same for both. Families have not been looked at with regard to progression. But, with regard to the type of pathology there is no difference. There are many candidate genes. The gene for NGF-converting enzyme which maps on chromsome 19 is a candidate gene. And there is a genetic study that suggests that there might be linkage to chromsome 19.

Van Broeckhoven
It has not been replicated by other groups. They used two approaches in the linkage study. They used the affected member pedigree analysis and the lod score analysis. With the affected pedigree member analysis they have a lod score of about 4, but with the lod score method the score drops to 2. There is no shift in the curves; just the lod score drops, which is a bit strange.

Barnard
Could you explain a little further about the proteolytic processing of the APP. You showed an age-dependent increase in resistance to this processing and at the same time you showed in Alzheimer specimens an increase in proteolytic products.

Beyreuther
We have different proteases that handle this molecule. The proteases that I would call the pathological proteases generate the amyloid as an insoluble breakdown product. And then we have other turnover processes. The major sites for APP processing are lysosomes that handle the amyloid precursor and then we have another APP processing on the cell surface. The latter converts the membrane protein into a secretory form. We do not know much about the secretory form, but, for instance, T cells or macrophages make a lot of these secretory forms. To our surprise we can isolate more of the full length of molecule from Alzheimer's brains than from control brains. So we were tempted to speculate that there is something wrong with the proteases. When we analyzed the sera of these patients to see whether there is something wrong with proteases, we obtained preliminary evidence for altered proteolysis of the amyloid precursor in Alzheimer's disease. We found reduced proteolysis in the plasma of Alzheimer patients as compared to controls. We also have evidence for proteases that are upregulated in Alzheimer plasma and that are able to rapidly digest APP. We also investigated whether there are brain proteases that deal specifically with this amyloid precursor and do the inactivation cut, i.e., generate the secretory forms of the precursor. David Small, Melbourne, purified a protease that copurifies with acetylcholinesterase. This protease does the right job, i.e., releases the precursor from the extracellular matrix (Small et al. Biochemistry 30:10795-10799, 1991).

Racagni
Do you think there might be a preferential association of the amyloid precursor with the specific neurons?

Beyreuther
This has not been found. But what is interesting in this context is that axotomy of peripheral nerves upregulates amyloid gene expression. Mainly the protease inhibi-

tor-containing form of the precursor increases. If regeneration is prevented this upregulation is maintained over 20 days. This suggests to us again that this molecule which has something to do with pathfinding and maintenance of the synaptic density is a "repair protein." If you do not complete the repair process the neuron keeps making the precursor. That seems to be the problem in Alzheimer's disease. Too much of this protein is made, which means too much amyloid precursor is present and hence a higher risk of developing Alzheimer's disease. That is the rationale behind all these regulatory studies. There is a clear association of the upregulation with axotomy and regeneration. But these investigations have been done in the peripheral nervous system and not yet in brain.

Vogel

You said that the amyloid pathology is always bilateral but never symmetrical. Is there a preference for any side, right or left side of the brain?

Beyreuther

In the right to left comparison there is a 10%–20% difference but there is no preference for any side. We should do a cohort study on this with familial cases. From the work on Down' patients one could postulate that the problem starts in the enthorhinal cortex and spreads from there. But even in Down's syndrome pathology is already so advanced that it is bilateral. But I want to remind you that as soon as we see pathology the process goes on for a minimum of 1 year.

Van Broeckhoven

In relation to memory deficits being an early sign of Alzheimer's disease, this is not mandatory. We are working with an Alzheimer familiy having very early onset of about 35 years. In one of the probands behavioral changes were the first sign instead of memory impairment. Within that same family another proband had apparently slight memory problems. In the same family we asked all the probands at age 35 if they would agree to have a postmortem examination if they were to die in an accident or of cancer and they all did. However, the chance of one of them dying in such a way is very low. In families with higher age at onset with siblings, you actually are not sure whether the parent or the sib was a genetic case or a sporadic patient within a family. Thus I agree, it is rather difficult to obtain suitable controls.

Beyreuther

The Down's data are very good. But we have to get data from Alzheimer families to finally convince ourselves and to prove that the amyloid pathology is not an epiphenomenon.

Molecular Genetics of Familial Alzheimer's Disease

C. Van Broeckhoven and J.-J. Martin

Introduction

The successful identification of the defective gene in a number of heritable diseases (i.e., Duchenne muscular dystrophy, neurofibromatosis, fragile X syndrome) using molecular genetic techniques has stimulated several scientists to use a similar approach to Alzheimer's disease (AD), the major cause of senile dementia in the Western population (Davies 1986).

Although sporadic AD occurs frequently, it has been recognized for many years that genetic factors play an important role in the etiology of AD. A large number of families have been reported in which AD seems to segregate according to an autosomal dominant disease with age-dependent expression. In these families a child of an affected parent has a 50% chance of inheriting a gene predisposing him or her to AD. It is this gene that molecular geneticists are interested in.

The approach used to localize and identify the gene underlying the pathology in an affected family is often referred to as the method of "reverse genetics." Basically this method aims at linking the disease gene to a polymorphic DNA marker of which the chromosomal location is already known. Cosegregation of a polymorphic DNA allele with the disease indicates that the latter must be located close to the DNA marker or, in other words, that the disease is genetically linked to the DNA marker. Once linkage is found, the chromosomal location of the disease gene is known and its subsequent identification depends on the use of a number of molecular genetic techniques (Wicking and Williamson 1991).

Chromosome 21

The first clue to where to look for an AD gene followed the observation of classical AD neuropathology in the brain of aged Down syndrome patients, i.e., senile plaques, neurofibrillary tangles, and congophilic angiopathy. The presence of three copies of chromosome 21 in Down syndrome patients led to the hypothesis that overexpression of a chromosome 21 gene may be responsible for the observed brain pathology and thus for AD. Therefore it did not come as much of a surprise that the first positive linkage results were indeed obtained with chromosome 21 DNA markers. In 1987, St. George-Hyslop et al. (1987) reported significant linkage results in four early onset (M < 60 years) AD families with DNA markers located on the proximal long arm of chromosome 21, suggesting that a familial AD (FAD) gene resides in this chromosomal region.

Almost simultaneously with the FAD linkage report, the gene coding for the amyloid precursor protein (APP), the substrate for the ß amyloid protein deposited in the senile plaques and cerebral vessels of the brains of AD patients, was localized to chromosome 21 close to the AD linked markers (Kang et al. 1987; Tanzi et al. 1987a). Initial excitement that the FAD gene may have been found disappeared very rapidly, however, following the observation of multiple recombinants between the disease and APP excluding the latter as the primary site of mutation in FAD (Van Broeckhoven et al. 1987; Tanzi et al. 1987b).

Genetic Heterogeneity

Several researchers tried to confirm the chromosome 21 linkage in their set of AD families, resulting in both positive (Goate et al. 1989) and negative (Pericak-Vance et al. 1988; Schellenberg et al. 1988) reports. The positive report contained information obtained in six early onset AD families while the negative linkage studies comprised mainly late onset (M > 60 years) AD families, suggesting that AD is genetically heterogeneous with the chromosome 21 FAD locus representing the genetic defect only in the early onset AD families.

In an international collaborative study including 48 families with both early and late onset of AD, St. George-Hyslop et al. (1990) obtained significant evidence proving again that chromosome 21 harbors an FAD gene. However, only some of the results from early onset AD families, but none of them from late onset AD families, suggested linkage to chromosome 21. Essentially the same observation was made by Schellenberg et al. (1991b) analyzing a different, nonoverlapping set of 48 AD families.

Pericak-Vance et al. (1991) examined DNA markers from other chromosomes and found suggestive linkage of FAD in late onset AD families with markers of the proximal region of the long arm of chromosome 19. Interestingly, the evidence was particularly strong when the linkage analysis was restricted to patients, excluding the at risk individuals who may or may not be gene carriers. Although this positive linkage finding warrants further investigation, independent confirmation of the chromosome 19 linkage is needed before the presence of an FAD gene in this region of chromosome 19 can be generally accepted.

Mutations in the APP Gene

The confirmation of the chromosome 21 linkage as well as the presence of genetic heterogeneity even in early onset familial AD reinforced the possibility that the APP gene may after all contain a mutation in some families. Recently, Goate et al. (1991) found cosegregation of a single base mutation in exon 17 of the APP gene with FAD in two early onset AD families. Both AD families demonstrated linkage of FAD with APP in the absence of recombinants. Although the APP gene constitutes 18 exons, with exons 16 and 17 coding for the ß amyloid, exon 17 was sequenced first since a mutation in this exon had been reported in families with hereditary cerebral hemorrhage with amyloidosis of Dutch type (HCHWA-D) (Van Broeckhoven et al. 1990a; Levy et al. 1990; Bakker et al. 1991). In the brain of HCHWA-D patients the same ß amyloid is deposited primarily in the cerebral vasculature, leading to massive intracerebral hemorrhages and subsequent death of the patients around the age of 45-65 years (Haan et al. 1991). In HCHWA-D the mutation replaces a glutamate with a glutamine at APP residue 693 close to the extracellular site of proteolytic breakdown of the ß amyloid. So, it has been hypothesized that the APP_{693} mutation affects the normal metabolism of APP in HCHWA-D patients. In AD the amino acid substitution is highly conservative, replacing an isoleucine for a valine at APP residue 717 located within the transmembrane domain of APP. It is not clear yet how the APP_{717} mutation may be responsible for AD.

The APP_{717} mutation was also observed in patients of two Japanese early onset AD families (Naruse et al. 1991). The absence of the APP_{717} mutation in normal individuals as well as its presence in AD patients of different ethnic background support the assumption that the APP_{717} mutation is pathogenic. Since the APP_{717} mutation was not found in a large number of either early or late onset AD patients, it clearly is not a common cause of AD (Van Duijn et al. 1991; Chartier-Harlin et al. 1991a; Schellenberg et al. 1991a). However, it cannot yet be excluded that in these patients other mutations may be present in the APP gene.

A Second FAD Locus on Chromosome 21

In some early onset AD families genetic linkage studies suggested an FAD gene located centromeric of APP, implying a second FAD gene on chromosome 21 nearer to the centromere (St. George-Hyslop et al. 1990). We have been examining two multigeneration AD families, AD/A and AD/B, in which the disease is inherited as an autosomal dominant trait. In both families the patients are very young at onset, with a mean age of 35 years. The clinical and neuropathological features of the patients have recently been reviewed and updated and are consistent with early onset AD (Martin et al. 1991). Since both families live close to each other, and since the patients in the families share the factor of early onset age, we believe that they may in fact be part of one single pedigree. However, genealogical studies have not been able to provide evidence that links the two families together.

In 1987, we excluded the APP gene as the site of mutation based on the observation of one inferred recombinant in family AD/A (Van Broeckhoven et al. 1987). Subsequently, linkage of families AD/A and AD/B with chromosome 21 markers suggested the presence of an FAD gene closer to the centromere (Van Broeckhoven et al. 1990b; St George-Hyslop et al. 1990). Today, we have multiple obligate recombinants with APP in both families excluding the FAD gene over a region of 15% recombination of the APP gene. Further, linkage with the pericentromeric chromosome 21 marker D21S13 is highly suggested, with a probability of linkage versus nonlinkage of 100 to 1. Other markers in the same region have either been uninformative or showed slightly negative results. So, the question remains of whether there is a second FAD gene on chromosome 21 in families AD/A and AD/B or if the APP recombinants are a spurious finding. The latter is highly unlikely since some of the APP recombinants occurred between histopathologically confirmed AD patients belonging to different branches in the pedigrees. Screening of the patients in these families for the APP_{717} mutation was negative (Chartier-Harlin et al. 1991a). In the collaborative study of St George-Hyslop et al. (1990) there was evidence in one other AD family suggestive of linkage with centromeric chromosome 21 markers in the presence of multiple APP recombinants, the Italian family FAD2. However, although the data indicate that the pericentromeric region of chromosome 21 is still a potential site for an FAD gene in some families, it needs to be interpreted with caution. The linkage in each family has not reached a level where it becomes significant, i.e., a linkage probability of 1000 to 1. Further linkage studies will be needed to evaluate the possibility of a second FAD locus on chromosome 21.

It is interesting that the chromosome 21 linkage has revived the hypothesis that AD may result from the accumulation of trisomic cells during the life of the individual (Potter 1991). The trisomy 21 cells develop over time by unequal chromosomal segregation during mitosis. The onset age depends on the percentage of trisomic cells, and AD results as does Down syndrome from overexpression of the APP gene and/or of other genes on chromosome 21. In light of this hypothesis the second chromosome 21 FAD locus may be interpreted as a genetic defect predisposing dividing cells to the trisomic stage.

Conclusions

Overall, genetic linkage studies have indicated that the molecular genetics of FAD is extremely complex. Evidence was provided for a genetic defect on chromosome 21 responsible for early onset FAD. However, not all early onset AD families are linked to chromosome 21. Further, in a few early onset families a rare mutation was detected in the APP gene, while in other families the genetic defect seems to be located closer to the centromere. Other genes must exist that predispose to AD particularly in the late onset AD families. Such a gene is possibly located on chromosome 19.

Note Added in Proof. Two other mutations have recently been identified in AD families, both involving the same codon 717 of APP and substituting the valine for a phenylalanine (Murrell et al. 1991) or a glycine (Chartier-Harlin et al. 1991b). It seems unlikely that finding different mutations involving the same codon is a coincidence. However, the exact mechanism by which mutations at this site cause the amyloid deposition in AD pathology remains to be elucidated.

References

Bakker E, Van Broeckhoven C, Haan J et al. (1991) DNA diagnosis for hereditary cerebral haemorrhage with amyloidosis (Dutch type). Am J Hum Genet 49:518-521

Chartier-Harlin M-C, Crawford F, Hamandi K et al. (1991a) Screening for the ß-amyloid precursor protein mutation (APP717:Val ---> Ile) in extended pedigrees with early onset Alzheimer's disease. Neurosci Lett 129:134-135

Chartier-Harlin M-C, Crawford F, Houlden H et al. (1991b) Early-onset Alzheimer's disease caused by mutations at codon 717 of the ß-amyloid precursor protein gene. Nature 353:844-846

Davies P (1986) Genetics of Alzheimer's disease: a review and a discussion of the implications. Neurobiol Aging 7:459-466

Goate AM, Owen MJ, James LA et al. (1989) Predisposing locus for Alzheimer's disease on chromosome 21. Lancet 1:352-355

Goate AM, Chartier-Harlin M-C, Mullan M et al. (1991) Segregation of a missense mutation in the amyloid precursor protein gene with familial Alzheimer's disease. Nature 349:704-706

Haan J, Hardy JA, Ross RAC (1991) Hereditary cerebral haemorrhage with amyloidosis - Dutch type: its importance for Alzheimer research. Trends Neurosci 14:231-234

Kang J, Lemaire H-G, Unterbeck A et al. (1987) The precursor of Alzheimer's disease amyloid A4 protein. Nature 325:733-736

Levy E, Carman MD, Fernandez-Madrid I et al. (1990) Mutation of the Alzheimer's disease amyloid gene in hereditary cerebral haemorrhage, Dutch type. Science 248:1124-1126

Martin J-J, Gheuens J, Bruyland M et al. (1991) Early-onset Alzheimer's disease in 2 large Belgian families. Neurology 41:62-68

Murrell J, Farlow M, Ghetti B, Benson D (1991) A mutation in the amyloid precursor protein associated with hereditary Alzheimer's disease. Science 254:97-99

Naruse S, Igarashi S, Kobayashi H et al. (1991) Mis-sense mutation Val ---> Ile in exon 17 of amyloid precursor protein gene in Japanese familial Alzheimer's disease. Lancet 337:979

Pericak-Vance MA, Yamaoka LH, Haynes CS et al. (1988) Genetic linkage in Alzheimer's disease families. Exp Neurol 102:271-279

Pericak-Vance MA, Bebout JL, Gaskell PC Jr et al. (1991) Linkage studies in familial Alzheimer disease: evidence for chromosome 19 linkage. Am J Hum Genet 49:1034-1050

Potter H (1991) Review and hypothesis: Alzheimer disease and Down syndrome - chromosome 21 non-disjunction may underlie both disorders. Am J Hum Genet 48:1192-1200

Schellenberg GD, Bird TD, Wijsman EM et al. (1988) Absence of linkage of chromosome 21q21 markers to familial Alzheimer's disease. Science 241:1507-1510

Schellenberg GD, Anderson L, O'Dahl S et al. (1991a) APP_{717}, APP_{693}, and PRIP gene mutations are rare in Alzheimer disease. Am J Hum Genet 49:511-517

Schellenberg GD, Pericak-Vance MA, Wijsman EM et al. (1991b) Linkage analysis of familial Alzheimer disease, using chromosome 21 markers. Am J Hum Genet 48:563-583

St. George-Hyslop PH, Tanzi RE, Polinsky RJ et al. (1987) The genetic defect causing familial Alzheimer's disease maps on chromosome 21. Science 253:885-890

St. George-Hyslop PH, Haines JL, Farrer LA et al. (1990) Genetic linkage studies suggest that Alzheimer's disease is not a single homogeneous disorder. Nature 347:194-197

Tanzi RE, Gusella JF, Watkins PC et al. (1987a) Amyloid ß protein gene: cDNA, mRNA distribution and genetic linkage near the Alzheimer locus. Science 235:880-884

Tanzi RE, St. George-Hyslop PH, Haines JL et al. (1987b) The genetic defect in familial Alzheimer's disease is not tightly linked to the amyloid ß-protein gene. Nature 329:156-157

Van Broeckhoven C, Genthe CA, Vandenberghe A et al. (1987) Failure of familial Alzheimer's disease to segregate with the A4 amyloid gene in several European families. Nature 329:153-155

Van Broeckhoven C, Haan J, Bakker E et al. (1990a) Amyloid beta protein precursor gene and hereditary cerebral haemorrhage with amyloidosis (Dutch). Science 248:1120-1122

Van Broeckhoven C, Backhovens H, Van Hul W et al. (1990b) Genetic linkage analysis in early-onset familial Alzheimer's disease. In: Bunney WE Jr, Hippius H, Laakmann G, Schmauss M (eds) Neuro-psychopharmacology. Springer, Berlin Heidelberg New York, pp 86-91

Van Duijn CM, Hendriks L, Cruts M et al. (1991) Amyloid precursor protein gene mutation in early-onset Alzheimer's disease. Lancet 337:978

Wicking C, Williamson R (1991) From linked marker to gene. Trends Genet 7:288-293

Discussion of Presentation C. Van Broeckhoven and J. J. Martin

Barnard

Can your findings be related to those of Dr. Sorbi with the Italian group, where all the families but two showed no linkage to chromosome 21. How many early onset families were in that population?

Sorbi

The family which showed the strongest evidence for linkage is the same as family FAD4 in the original linkage study of Peter St. George-Hyslop (1987). All families in the Italian study were early onset families.

Van Broeckhoven

Today there are two AD families known that show a lod score of 2 (probability of linkage of the family is 100:1), namely, family FAD4 in the Italian study and family AD/A in the Belgian study. The other families are not really informative. Up to now people have been using classical RFLPs in the linkage studies. However, most of these markers detect diallelic polymorphisms. Therefore, these markers are not really useful in Alzheimer pedigrees since in most cases the patients are dead and their genetic information can only be reconstructed relying on data obtained in their children. All the uninformative families need to be reanalyzed with highly informative markers such as $(CA)_n$ repeat polymorphisms before we can say anything about the actual proportion of chromosome 21 linked families.

Further, I would like to quote Margaret Pericak-Vance (Durham, USA) and Jonathan Haines (Boston, USA) who claim that they have not been able to exclude chromosome 21 completely in some of the late onset AD families.

Reich

At this point I would assume there are five forms of Alzheimer's disease: two linked to chromosome 21, one on 19, familial Alzheimer's that is not on 21 or 19, and nonfamilial Alzheimer's.

Van Broeckhoven

We should not forget that the chromosome 19 linkage has not been replicated yet. Thus I suggest four forms of AD. But, I would also be very careful of saying that there are families not linked to 19 and 21, because it is still possible that these families are simply not informative and therefore generating either slightly positive or slightly negative lod scores. If you have a sufficient number of these uninformative families you may eventually exclude linkage. Family AD/B in our study, for example, generates slightly negative lod scores since the marker D21S13 detects an interbranch recombinant but is not informative in the rest of the pedigree. Unless all the AD families have been analyzed with highly informative markers, I would not make any statements about the number of possible AD loci.

Peltonen

Could one of the linkage experts explain why a significant lod score obtained in the chromosome 19 linkage "affected only" drops to almost nonsignificant if in the linkage analysis the unaffected individuals are also included.

Van Broeckhoven

In the linkage study "affected only" only the affected were included. The lod score obtained was 4. If the unaffected are included in the linkage analysis the lod score drops to nearly 2. The most likely explanation is that the affected and unaffected share the same haplotype. If the unaffected individual is a true escapee, the sharing of the haplotype leads to a recombinant. If, however, the unaffected is a nonpenetrant gene carrier, sharing the haplotype leads to a false recombinant. In both cases the lod score will be significantly lowered.

Analysis of Psychiatric Disease by Molecular Genetics*

A. Poustka and P. Kioschis

Introduction

Due to the technical developments in molecular biology and molecular genetics there has been progress from being able to analyze genes with known gene defect (e.g., HPRT, PKU) to being able to analyze genes with unknown gene defect localizable using genetic or cytogenetic techniques (e.g., DMD, CF). The newest stage of this development, the human genome project, holds promise of being an efficient route by which to identify many or all of the genes of the human genome and is of special relevance in diseases with more complex genetics, including psychiatric disorders.

Since the genome contains the information underlying the development and function of the entire organism, it is not unexpected that most diseases, including many psychiatric disorders, have strong genetic components. The first group of diseases in which the mechanism of action could be demonstrated belonged to a group in which the defective gene product could be directly identified (e.g., phenylketonuria, thallasemias). The current focus of the work in human genetics is on diseases in which the gene defect is not known, but in which the location of the gene can be followed by either family analysis or, in fortunate cases, also by the cytogenetic identification of chromosomal rearrangements (translocations or deletions), which can be localized more easily. This information can then be used to identify the gene and to understand the defect in molecular terms. After the gene has been isolated, the structure and often the function of the affected gene product can be identified and studied, as is possible for the class of diseases with known gene product, with the hope that an understanding of the affected gene product and its mechanism of action will allow the development of tools for diagnosis and treatment of the disease. As will be described later using the example of the fragile X mutation, this type of approach has led to the identification of a number of genes involved in genetic diseases over the past few years (Monaco et al. 1986; Rommens et al. 1989; Riordan et al. 1989; Gessler et al. 1990; Herrmann et al. 1990). The technology,

*This work was supported by a grant from the Deutsche Forschungsgemeinschaft and by the Human Frontier Science program.

which is still very new, now has to be extended to the analysis of the class of diseases with complex genetics. As described in the last part of this paper, it is, however, likely that the rapid progress in the global analysis of the human genome will contribute greatly to the identification of genes involved in many common diseases with complex genetics, a group which includes many of the psychiatric diseases. As example of this type of analysis, the global analysis of the human Xq28 region will be described, a 9-megabase region extending from the fragile X locus to the telomere of the long arm of the human X chromosome which has been found to be involved in a large number of inherited human disorders (Mandel et al. 1989; Mc Kusick 1989).

The Molecular Analysis of the Fragile X Mutation

Fragile X, the most common inherited form of mental retardation, is characterized cytogenetically by the appearance of a folate-sensitive fragile site in Xq27.3 (Sutherland 1977). The disease is also characterized by an extremely unusual pattern of inheritance, with a 20% frequency of males transmitting the disease without showing the fragile X phenotype (mental retardation, unusual facial features, macroorchidism) (Sherman et al. 1985; Nussbaum and Ledbetter 1986).

Since the fragile site observed cytogenetically also leads to enhanced breakage of chromosomes during cell line construction (Warren et al. 1987), the group of Steve Warren has been able to construct a set of human-hamster cell hybrids containing the region from the fragile site to the telomere, the only human DNA which has been translocated onto a hamster chromosome (Warren et al. 1990). To localize the position of the break point in the cell hybrid expected to be close to the position of the fragile site, we therefore established pulsed field gel restriction maps from the entire Xq28 area. For this work the DNA of a cell hybrid containing an intact X chromosome (Wieacker et al. 1984) and the DNA of a number of hybrids expected to be broken at the position of the fragile site was used. Comparison of these maps, extending over a region of 12 Mb, showed only few differences throughout the area and allowed us to localize the position of the breakpoint 9 Mb proximal of the telomere, within an area of 300 kb, at which the two maps diverged completely (Poustka et al. 1991; Dietrich et al. 1992).

Analysis of pulsed field gel digests of DNA of patients and normal controls, using methylation-sensitive enzymes, identified another feature associated with the fragile X mutation, the selective methylation of a cluster of rare cutter sites close to the position of the cell line breakpoint in the DNA of patients expressing the fragile X phenotype (Vincent et al. 1991; Bell et al. 1991; Dietrich et al. 1991).

As the next step in the analysis of the mutation, yeast artificial chromosome (YAC) clones were isolated from this area, resulting in a YAC contig of 2 Mb in size spanning the region of the translocation breakpoints in the cell lines (Dietrich et al. 1991). The immediate area surrounding the translocation was then isolated in cosmid clones. Expecting larger rearrangements in the vicinity of the fragile region, cosmids were isolated from a library derived from a normal X chromosome (Nizetic et al. 1991) and in parallel from a library constructed from a patient chromosome. To be able to close a gap resulting in a cosmid contig of 300 kb spanning the entire region, an additional cosmid library had to be constructed from one of the YAC clones. The analysis of the cosmid and YAC clones allowed then the localization of the position of the majority of cell line breakpoints and the selectively methylated rare cutter restriction enzyme sites close to each other within a 40-kb interval (Dietrich et al. 1991). By using cosmids flanking the cluster of rare cutter sites for in situ hybridization on patient chromosomes induced to show the fragile site, we were able to map the cytogenetically observed fragile site to the same interval, demonstrating the close proximity of the fragile site, the breakpoints in the cell hybrids, and the CpG island — CG rich regions in the genome found in association with gene promoters — showing selective methylation in fragile X patients.

In this area containing the CpG island and the cluster of breakpoints we were also looking for a sequence which could be the mutation responsible for the expression of the fragile site. Using the EcoR1 fragment containing the CpG island as well as the breakpoint cluster as a probe on Southern blots, this region could be shown to contain a segment showing different size in normal controls (characterized by an EcoRI fragment of 5.1 kb), in phenotypically normal transmitting males (characterized by a 5.2- to 5.5-kb fragment) and in patients (identified by an up to 7-kb fragment, also recognizing a smear indicating somatic heterogeneity). This size difference could be shown (Yu et al. 1991) to be caused by the variation in the size of a CGG repeat sequence, varying in size from approximately 30 repeats in normal controls, over approximately 70 copies in transmitting males or females, to a few hundred in fragile X patients.

To isolate genes involved in the expression of the fragile X phenotype, we screened a human fetal brain cDNA library with cosmid clones and unique fragments from this area, leading to the identification of two cDNA sequences. One of these (FMR1), also identified in another laboratory (Verkerk et al. 1991), could be shown to contain the CGG repeat in its first exon. It is therefore very likely that this gene is responsible for the phenotype of the fragile X mutation. The cDNA sequence is well conserved (sequence homology can be observed down to nematode and yeast), is expressed mainly in brain and less in other tissues, and could be shown to be uniformly expressed in all cell types of the rat brain by in situ hybridization, which leads us to speculate that FMR1 is a brain household gene (A. Poustka and P. Seeburg, unpublished).

Since the cytogenetic localization of the fragile X mutation has been known for more than 20 years (Lubs 1969) and later genetic analysis was able to localize the mutation in a region of a few CM (Oberle 1987), roughly corresponding to 5 Mb, the isolation of this gene has been relatively straightforward using the newly developed technologies such as pulsed field gel electrophoresis and large insert cloning. The analysis of genes involved in psychiatric disorders is expected to be more complicated and will be more accessible using techniques which are currently being developed for the global analysis of the genome, the goal being to analyze all genes in a region (Poustka et al. 1986; Lehrach et al. 1990).

The Molecular Genetic Analysis of the Xq28 Region: Progress Towards an Integrated Clone and Gene Map

To extend this type of work to the large number of genes for genetic diseases located within the Xq28 region, we are currently carrying out a global analysis of this region of the human X chromosome and are expecting to combine information on the genetic map of the region, the physical mapping information mentioned above, information on YAC and cosmid contigs, and information on transcripts in this area (Fig. 1).

The first step in this analysis was to complete the physical map of the entire area. We are now in the process of constructing ordered YAC and cosmid clones and have started to identify as many as possible of the coding regions located in this area to give access to some of the disease genes in Xq28. Since this region is very dense in genes and in genetically and physically mapped markers, it is ideal as a model system to attempt to get an integrated clone and gene map of a region.

As part of this process, we have constructed cosmid and linking clone libraries (in cosmid vectors) from Q1Z, the cell hybrid containing Xq28 in a hamster background. Human clones were identified by hybridization with human repeat sequences, and clones corresponding to a sixfold coverage of the region in cosmids and a 20-fold coverage of the region in NotI linking clones were picked into microtiter plates and used to construct high density filter grids (A. Poustka, unpublished), using a robot developed at the Imperial Cancer Research Fund (ICRF). In parallel, we are in the process of identifying YAC clones from this region, using unique or pooled probes to screen high density filter grids of a YAC clone library constructed at the ICRF (Larin et al. 1991). These clone grids are being screened with

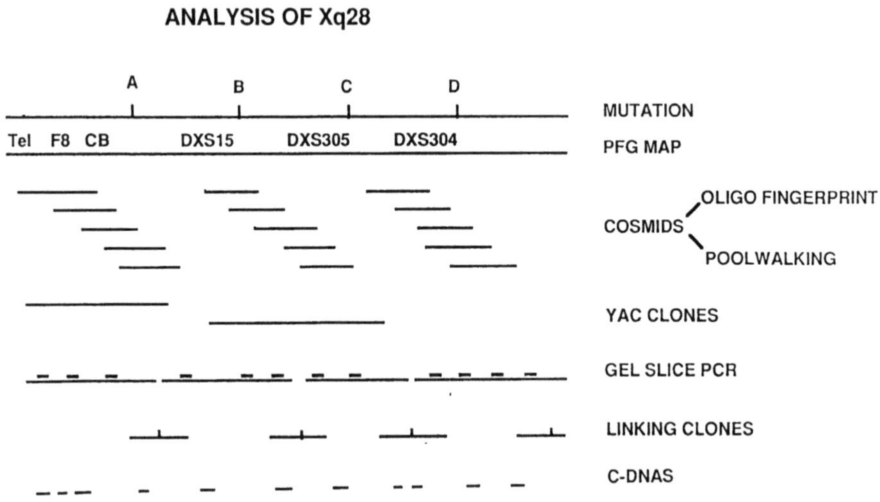

Fig. 1. Analysis of Xq28. The relationship of different types of biological information and of different types of probes to the ordered cosmid clones. *PCR*, polymerase chain reaction; *Tel*, telomere; *F8*, factor VIII; *CB*, color blindness

a number of different probe types expected to provide information on clone overlaps (hybridization with oligonucleotides and pools of cosmids), correspondences between different clone types (hybridization of cosmids with YACs and YACs with cosmids), with the physical map (hybridization of clones with polymerase chain reaction products from bands from pulsed field gels, with unique probes localized within the map) and genes (hybridization of clones with radiolabeled cDNAs) (Poustka 1990; Poustka et al. 1986; Lehrach et al. 1990). We are currently analyzing a large number of hybridizations, a process which has already led to the identification of a number of extended cosmid contigs spreading over this region (contigs up to 400 kb in length are already identified). This data set as well as new data currently being generated will be used in collaboration with the group of Hans Lehrach (ICRF, London) to develop a data base integrating all the information available on this region.

We expect that this work will result in the establishment of an integrated physical, genetic, clone and transcript map of Xq28, provide simplified access to candidate genes for diseases from the region, and serve as a test for the identification and analysis of genetic information from larger regions of the human genome. This is likely an essential step in the identification of genes involved in some of the psychiatric disorders.

References

Bell MV, Hirst MC, Nakahori Y, MacKinnon RN, Roche A, Flint TJ, Jacobs PA, Tommerup N, Tranebjaerg L, Froster-Iskenius U, Kerr B, Turner G, Lindenbaum RH, Winter R, Penbrey M, Thibodeau S, Davies KE (1991) Physical mapping across the fragile X: hypermethylation and clinical expression of the fragile X syndrome. Cell 64:861 866

Dietrich A, Kioschis P, Monaco AP, Gross B, Korn B, Williams S, Sheer D, Heitz D, Oberlé I, Toniolo D, Warren ST, Lehrach H, Poustka A (1991) Molecular cloning and analysis of the fragile X region in man. Nucleic Acids Res 19:2567-2572

Dietrich A, Korn B, Poustka A (1992) Completion of the physical map of Xq28: the location of the gene for L1CAM on the human X chromosome. Mammalian Genome (in press)

Gessler M, Poustka A, Cavenee W, Neve RL, Orkin SH, Bruns GAP (1990) Homozygous deletion in Wilms tumour of a zinc-finger gene identified by chromosome jumping. Nature 343:774

Herrmann BG, Labeit S, Poustka A, King TR, Lehrach H (1990) Cloning of the T gene required in mesoderm formation in the mouse. Nature 343:617

Larin Z, Monaco AP, Lehrach H (1991) Yeast artificial chromosome libraries containing large inserts from mouse and human DNA. Proc Natl Acad Sci USA 88:4123-4127

Lehrach H, Drmanac R, Hoheisel J, Larin Z, Lennon G, Monaco AP, Nizetic D, Zehetner G, Poustka A (1990) Hybridization fingerprinting in genome mapping and sequencing. In: Davies KE, Tilghman S (eds) Cold Spring Harbor Laboratory, Cold Spring Harbor, pp 39-81 (Genome analysis, vol 1)

Lubs HA (1969) A marker X chromosome. Am J Hum Genet 21:231-244

Mandel JL, Willard HF, Nussbaum RL, Romeo G, Puck JM, Davies KE (1989) Report of the committee on the genetic constitution of the X chromosome. Cytogenet Cell Genet 51:384-437

McKusick VA (1989) Mendelian inheritance in man. Catalogs of autosomal dominant, autosomal recessive, and X-linked phenotypes, 9th edn. John Hopkins University Press, Baltimore

Monaco AP, Neve R, Colletti-Feener C, Bertelson CJ, Kurnit DM, Kunkel LM (1986) Isolation of candidate cDNAs for portions of the Duchenne muscular dystrophy gene. Nature 323:646

Nizetic D, Zehetner G, Monaco AP, Young BD, Lehrach H (1991) Construction, arraying, and high-density screening of large insert libraries of human chromosome X and 21: their potential use as reference libraries. Proc Natl Acad Sci USA 88:3233-3237

Nussbaum RJ, Ledbetter DH (1986) Fragile X syndrome: a unique mutation in man. Annu Rev Genet 20: 109-145

Oberle I, Camerino G, Wrogemann K, Arveiler B, Hanauer A, Raimondi E, Mandel JL (1987) Multipoint genetic mapping of the Xq26-28 region in families with the fragile X mental retardation and in normal families reveals tight linkage of markers in q26-27. J Hum Genet 77:60-65

Poustka A (1990) Physical mapping by PFGE. METHODS: a companion to Methods Enzymol 1:2, 204-211

Poustka A, Pohl T, Barlow DP, Zehetner G, Craig A, Michiels F, Ehrich E, Frischauf AM, Lehrach H (1986) Molecular approaches to mammalian genetics. Cold Spring Harb Symp Quant Biol LI:131

Poustka A, Dietrich A, Langenstein G, Toniolo D, Warren ST, Lehrach H (1991) Physical map of Xq27-Xqter: localizing the region of the fragile X mutation. Proc Natl Acad Sci USA 88:8302-8306

Riordan JR, Rommens JM, Kerem BS, Alon N, Rozmahel R, Grzelczak Z, Zielenski J, Lok S, Plavsic N, Chou JL, Drumm ML, Iannuzzi MC, Collins FS, Tsui LP (1989) Identification of the cystic fibrosis gene: cloning and characterisation of complementary DNA. Science 245:1066

Rommens JM, Iannuzzi MC, Kerem B-S, Drumm ML, Melmer G, Dean M, Rozmahel R, Cole JL, Kennedy D, Hidaka N, Zsiga M, Buchwald M, Riordan JR, Tsui L-P, Collins FS (1989) Identification of the cystic fibrosis gene: chromosome walking and jumping. Science 245:1059

Sherman SJ, Jacobs PA, Morton NE, Froster Isckenius U, Howard Peebles PN, Nielsen KB, Partington MW, Sutherland GR, Turner G, Watson M (1985) Further segregation analysis of the fragile X syndrome with special reference to transmitting males. Hum Genet 69:289-299

Sutherland GR (1977) Fragile sites on human chromosomes: demonstration of their dependence on the type of tissue culture medium. Science 197:265-266

Verkerk AJMH, Pieretti M, Sutcliffe JS, Fu YH, Kuhl PDA et al. (1991) Identification of a gene (FMR1) containing a CGG repeat coincident with a breakpoint cluster region exhibiting length variation in fragile X syndrome. Cell 65:905-914

Vincent A, Heitz D, Petit C, Kretz C, Oberlé I, Mandel JL (1991) Abnormal pattern detected in fragile X patients by pulsed-field gel electrophoresis. Nature 24349:624-626

Warren ST, Zhang F, Licameli GR, Peters JF (1987) The fragile site in the somatic cell hybrids: an approach for molecular cloning of fragile sites. Science 237: 420-423

Warren ST, Knight SL, Peters JF, Stayton CL, Consalez GG, Zhang F (1990) Isolation of the human chromosomal band Xq28 within somatic cell hybrids by fragile site breakage. Proc Natl Acad Sci USA 87: 3856-3860

Wieacker P, Davies KE, Cooke HJ, Pearson PL, Williamson R, Bhattacharya S, Zimmer J, Ropers H-H (1984) Toward a complete linkage map of the human X chromosome: regional assignment of 16 cloned single-copy DNA sequences employing a panel of somatic cell hybrids. Am J Hum Genet 36:265-268

Yu S, Pritchard M, Kremer E, Lynch M, Nancarrow J, Baker E, Holman K, Mulley JC, Warren ST, Schlessinger D, Sutherland GR, Richards RI (1991) Fragile X genotype characterized by an unstable region of DNA. Science 252:1179-1181

Discussion of Presentation A. Poustka and P. Kioschis

Gurling
Could you just go over the argument you have for saying that not all individuals with fragile X have a mutation in the cgg region. The second question is, is it possible to identify the repeat fragment in the cgg region using PCR spanning the polyarginine?

Poustka
To your last question, PCR is actually almost impossible. We have not been able to get the right PCR products over the cgg repeat. We can get a PCR product, but it is instable and loses cgg repeats. The product we get by PCR is different in size and in patients we cannot get any product at all. Referring to the first question, most patients show the altered fragment and do not express the gene, but there are patients who show the altered fragment and, also, expression of the gene as has been shown by Pieretti et al. And there are patients who express the fragile site, who are mentally retarded but do not have an altered fragment; this could be due to another mutation.

Vogel
You know that there are about 20% of male patients with fragile X syndrome who are mentally normal. I have heard this might be due to the fact that the gene has to break down gradually over one or two generations and that before this happens, the mental retardation does not occur. Can you comment on these nonmanifesting cases of fragile X syndrome?

Poustka
We have not investigated enough patients yet. Taking together the data produced in several laboratories, normal transmitting males do carry the premutation, the slightly bigger fragment. They transmit it to their daughters, who are normal. The sons and daughters of these mothers then show the enlarged fragment and are mentally retarded. At the moment it looks as if, again, there are many exceptions or our data are not precise enough.

Barnard
I would like to comment that the Xq28 has been thought attractive as a region for genome sequencing universally: I know of at least three groups which are also pursuing the same operations. So, I wonder if the Human Genome Project Organization should not be integrating these efforts, since it seems to me wasteful for several laboratories using all of this technology to look at the same piece of DNA.

Poustka
I have good contact to almost all groups working in the Xq28 region. At the moment nobody is doing anything very similar to what I do. There is intense collaboration among the groups with exchange not only of ideas but also materials and technology.

Barnard
You said that now you could correlate the physical map with the genetic map. The physical map one can well understand, because you have got a set of cloned regions for it. But I think it is interesting to ask why there was the discrepancy with the genetic map and how you resolved it. For example, we were involved in getting the last version of the mouse order genes in Xq28 and that is the order which is now up in man. But until the very recent publications, the human gene mapping was showing different orders. When you correlated the physical and genetic map, how did you get the new genetic map?

Poustka
The new genetic map does not exist, because the order of the genetic and physical maps must be the same. But since this is now the new order, the linkage data will be adjusted to the new physical order.

Barnard
What this really means is the genetic map was inaccurate. You got it from the physical map but we do not really know why there were discrepancies in the genetic map.

Poustka
That is quite normal for genetic maps. We know that the resolution of linkage analysis is very good to localize genes, but in most cases does not tell us the precise order. Since there are different likelihoods for different orders, the genetic map reflects the most likely order. Linkage maps are based on recombination, which is not linear, and on probability calculations.

Mendlewicz
I would like to get back to the question raised by Dr. Vogel on mental retardation. In the recent work of Mandel's group in Strasbourg and Sutherland in Australia, there was some correlation between a methyl group in the DNA segment and the prevalence of mental retardation, even in males. Could you say something about that?

Poustka

No, I think at the moment we are in a kind of state where we are very close to resolving all these questions, but we do not have enough data. We know that the methylation is correlated with the expression of the gene. It seems as if when there is complete methylation, there is no expression of FMR1. If methylation is only partial, there is expression of the FMR1 gene. Still, we have not analyzed enough patients to be conclusive about the correlation between the methylation, expression of the gene, and the mental retardation.

Molecular Genetics in Neurodegenerative Disorders

K. Mikoshiba

The nervous system is composed of mainly two types of cells, neurons and glial cells. Neurons process information by forming networks. Glials cells play an important role in supporting neuronal function by secreting various cytokines or assisting saltatory conduction by forming myelin. They maintain close contact with each other in physiological processes in the nervous system. Many neuropathological mutants which are responsible for abnormalities in the development of neurons or glial cells, and causing abnormal behavior, have been reported. We have analyzed various kinds of mutants and present in this article the results of analyses of two types of mutants producing abnormalities in myelination and cerebellar development.

Myelin-Deficient Mutants

Glial cells in the central nervous system (CNS) are composed of oligodendrocytes, astrocytes, and microglia. Oligodendrocytes form multilamellar organelles, or myelin, which surrounds axons and serves as an insulator to facilitate the conduction of impulses (Morell 1984). A number of mutant mouse strains show dys- or demyelination in the CNS. When the myelin protein is lost, the lamellar structure of myelin loosens and therefore the function of myelin becomes poor, resulting in the shivering behavior (Campagnoni and Macklin 1988; Ikenaka et al. 1991; Mikoshiba et al. 1991). Hence, myelin is useful in correlating gene expression, morphological change, and behavioral changes. The two allelic mutants shiverer and myelin deficient are characterized by abnormal expression of the myelin basic protein (MBP) gene. MBP makes up about 30% of total myelin protein and is thought to be located at the major dense line of myelin, which is formed by the fusion of the cytoplasmic surfaces of the oligodendrocyte plasma membrane during myelination. Absence of MBP causes the absence of the major dense line of myelin structure. Studies of the mutants have provided many insights into the function of the molecular organization of the gene and expression of MBP.

Shiverer Mutant Mouse

The shiverer is an autosomal recessive mutant. The absence of MBP in the CNS was reported by immunohistochemical techniques using MBP antibody. Gel electrophoretic analysis using sodium dodecyl sulfate polyacrylamide gel electrophoresis

(SDS-PAGE) revealed the absence of MBP in all subcellular fractions from adult shiverer CNS.

Chimeric Analysis of the Mutant

In order to learn whether the absence of MBP is a primary or secondary phenomenon, chimera mice were produced by the aggregation of eight-cell-stage embryos from wild-type control and shiverer mutant mice. Immunohistochemical study revealed MBP-positive and MBP-negative myelinated sites mixed in the white matter of the chimera brain, suggesting that humoral factors do not cause the abnormality and that the poor myelin formation in shiverer mice is intrinsic to the cells themselves. Since shiverer-type myelin was found adjacent to normal myelin on the same axon, it was concluded that the cause of dysmyelination resides primarily in the oligodendrocytes themselves, and not in the neuron. Then it became possible to correlate the absence of MBP to the abnormality of MBP-synthesizing cell, the oligodendrocyte. Therefore, the molecular genetic study was applied, focusing on MBP gene expression.

Deletion of MBP Gene in the Shiverer

MBP mRNA was almost totally absent in northern blot studies. Southern blot and gene walking data have led to the conclusion that a large portion of the MBP gene is deleted. The deleted portion is about 20 kb, including exons 3-7 among the 32 kb of the MBP gene. Since normal MBP genes transferred to the shiverer by producing transgenic mice eliminated the shivering symptoms and morphological and biochemical abnormalities, the absence of MBP in shiverer is caused by deletion of the MBP gene (Fig. 1).

Myelin-Deficient Allelic Mutant to Shiverer

The MBP content in the CNS of myelin-deficient mice was greatly decreased, to 3%–5% of the control level. However, MBP was clearly detected in the CNS of myelin-deficient mice both by immunohistochemistry and SDS-PAGE. Since MBP was partially expressed here, it was expected that the mechanism of suppression of MBP gene expression was known. The size of MBP mRNA detected in the myelin-deficient mice was the same as the wild control although the MBP mRNA content was about 3% of the control level.

Duplication and Inversion of MBP Genes in Myelin-Deficient Mutants

MBP gene was found to be duplicated in myelin-deficient mice. The promoter region of the MBP gene was normal with regard to the sequence and in vitro promoter

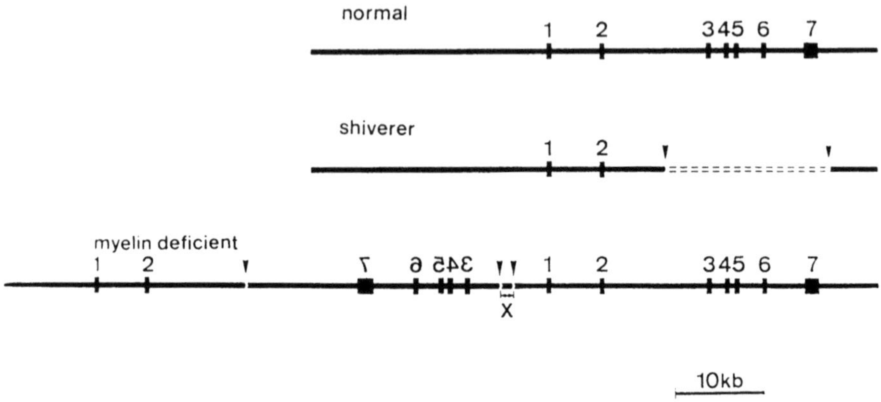

Fig. 1. Gene organization of myelin basic protein (MBP) gene in normal, shiverer, and myelin-deficient (*mld*) mutant mice. *Boxes* with the *number on top* show exons and their numbers. *Dotted line* with an *arrowhead* at each side shows where the gene is deleted in shiverer. *Arrowheads* in myelin-deficient MPB gene indicate the recombination points. Since MBP gene is duplicated, and a portion (including exon 3 to 7) of the upper gene is inverted, antisense RNA is produced from the inverted region, which is illustrated with *inverted numbers on the top* of exons (Mikoshiba et al. 1991)

activity. In situ chromosomal mapping of the MBP gene in these mice showed that the gene is located at the distal end of chromosome 18, indicating that the duplicated genes are both positioned very closely on the gene. By cosmid walking and Southern blot study it was found that a large upstream portion of the duplicated genes are inverted. The downstream portion was an intact copy.

Endogenous Antisense RNA Transcript

Although there is an intact gene in the duplicated MBP genes, the MBP gene expression is low. Therefore, it was predicted that antisense RNA could be synthesized from the inverted gene. The presence of antisense RNA in myelin-deficient mice was confirmed by nuclear run-on assay and RNase protection study. This is the first time anti-sense RNA has been transcribed in the same direction from a different region of the same gene in a vertebrate animal. RNA-RNA duplex has been detected in the brain of myelin-deficient mice.

Cerebellar Ataxic Mutant — Purkinje Cell-Deficient Mutants

The cerebellum plays an important role in motor function. There are only five types of neurons in the cerebellar cortex. Each neuron can be easily identified morphologically and the neuronal network in the cerebellum is well established. Among the cerebellar neurons, Purkinje cells play a key role in information processing. Purkinje cells receive more than 20 000 inputs from outside the cell, and the only output from the cerebellar cortex is the axon of the cell. There are many cerebellar ataxic mutants. Nervous and Purkinje cell degeneration mutants are Purkinje cell-deficient mutants.

P_{400} Protein in the Cerebellum of Purkinje Cell-Deficient Mutant Mice

Analysis of the cerebellum from Purkinje cell-deficient mutants detected that the high molecular weight membrane protein P_{400} was greatly decreased (Mikoshiba et al. 1979). The apparent molecular weight of P_{400} is 250 kDa as assessed by SDS-PAGE. P_{400} was found to be highly phosphorylated by cAMP-dependent protein kinase and also phosphorylated by calmodulin-dependent protein kinase and C kinase. Using ß-endo-*N*-acetylglucosaminidase F to digest P_{400} revealed that the protein has a small number of asparagine-linked oligosaccharide chains, and at immunohistochemical study three monoclonal antibodies stained Purkinje cells (Maeda et al. 1989; Nakanishi et al. 1991). Developmental expression by western blotting and immunohistochemical study showed that the time schedule of the expression of P_{400} corresponded closely to the morphogenesis and functional activity of Purkinje cells (Maeda et al. 1979; Nakanishi et al. 1991).

P_{400} as an Inositol 1,4,5-Trisphosphate Receptor

Phosphoinositide turnover system has recently been found to play an important role in signal transduction (Berridge and Irvine 1989). A specific receptor on the plasmalemma in response to extracellular signals couples to the G protein which activates phospholipase C. The activated phospholipase C hydrolyzes phosphatidylinositol 4,5 bisphosphate to produce 1,4,5-inositol-trisphosphate (InsP3). InsP3 has now become the second messenger known to release calcium from the intracellular calcium store. When isotope labeled InsP3 binding was performed on the cerebellar sections, InsP3 binding activity was the highest at the Purkinje cell body and the molecular layer where the dendrites of Purkinje cells are situated. But a very low amount of InsP3 binding activity was observed in the Purkinje cell-deficient cerebellum where a very slight P_{400} immunohistochemical reaction was observed (Maeda et al. 1989; Nakanishi et al. 1991). Recently Maeda et al. (1990) using three monoclonal antibodies determined that P_{400} protein is identical to InsP3 binding protein (Maeda et al. 1990; Supattapone et al. 1988; Fig. 2).

The cDNA libraries constructed by priming with random hexomer or oligo (dT) in the phage-λgt_{11} expression vector were screened with the three antibodies. From the amino acid sequence deduced from the nucleotide sequence of P_{400}, strong homology with the ryanodine receptor was observed. Ryanodine receptor is exclusively expressed in the sarcoplasmic reticulum of skeletal muscle. In situ hybridization has revealed that it is expressed on the cell body of Purkinje cells. Furuichi et al. (1989) constructed an InsP3 receptor expression plasmid by cloning the full cDNA into a vector that carries the ß-actinin promoter. The transfected cells increased the amount

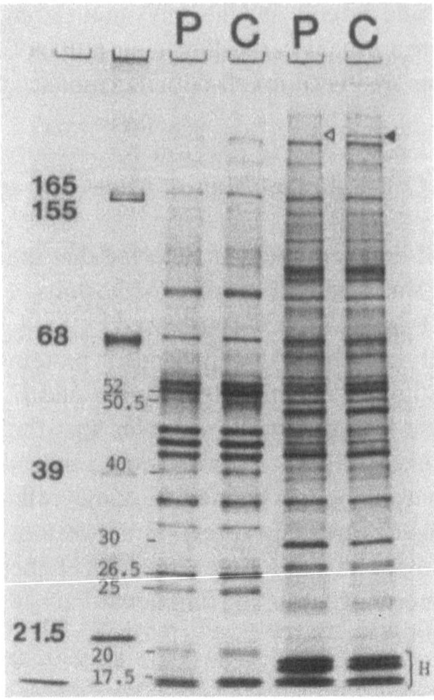

Fig. 2. Protein profile of the cerebellum from normal (*C*) and Purkinje cell-deficient mutant (*P*) mice. *Left two columns* are soluble protein fraction, and *right two columns* are from membrane fraction. *Closed arrowhead* indicates the protein P_{400}, which was finally proved to be InsP3 receptor. P_{400} protein is greatly decreased in Purkinje cell-deficient mutant, indicated by *open arrowhed. H*, histone bands. *Numbers* indicate the molecular weight (in kilodaltons)

of InsP3 binding and calcium-releasing activity according to increases in the amount of protein. Reconstitution of the purified InsP3 receptor into the liposome and planar lipid bilayer showed that InsP3 induced calcium release activity. Studies using immunogold techniques have revealed that InsP3 binding protein is highly enriched in the smooth surfaced endoplasmic reticulum (sER), but is found in very low amounts in Golgi apparatus, nuclei, mitochondria, and plasma membrane. Cross-linking experiments showed that InsP3 receptor forms homotetramer. The protein, greatly decreased in the cerebellum of Purkinje cell-deficient mutant mice, is a calcium release channel located in the sER. P_{400}/InsP3 binding protein is a good marker for the Purkinje cell and therefore can be used as a probe to diagnose the pathogenetic condition of the cerebellar ataxia.

Conclusion

In the first part of this paper I provided an example of myelin-deficient mutation. It is clear how gene expression is correlated with cell function and behavior. Analysis of the mutation by producing chimera animals is very important, since it enables us to analyze the events occurring between cells. The chimera between normal and diseased animals gives us information about which cell is primarily abnormal among the many types of cells in the brain. We can even manipulate the gene by producing transgenic mice to rescue the abnormal morphology and finally reverse the behavior to normal. The mouse which produces endogenous antisense RNA suppresses normal gene expression by forming RNA-RNA duplex produced from the sense

RNA transcribed from the normal gene. The molecular genetic analysis has made clear that there is a great difference at the gene level even though the symptoms or pathology are similar. With these ideas, it may be possible to introduce a new entity into the pathological classification already established. In the second part of my talk I showed an example of analysis of the Purkinje cell-deficient mutant which has led to the great discovery of the inositol trisphosphate receptor, a calcium releasing channel located on the sER — apart from the one on the plasma membrane. The antibody is now in good use as a probe to study cerebellar disease and the sCNA probe to get a pedigree of genetic disease of cerebellar ataxia.

References

Berridge MJ, Irvine R (1989) Inositol phosphates and cell signalling. Nature 341:197-205

Campagnoni AT, Macklin WB (1988) Cellular and molecular aspects of myelin protein gene expression. Mol Neurobiol 2:41-89

Furuichi T, Yoshikawa S, Miyawaki A, Wada K, Maeda N, Mikoshiba K (1989) Primary structure and functional expression of the inositol 1,4,5-trisphosphate-binding protein P_{400}. Nature 342:32-38

Ikenaka K, Okano H, Tamura T, Mikoshiba K (1991) Recent advances in studies on genes for myelin proteins. Dev Growth Differ 33:181-192

Maeda N, Niinobe M, Inoue Y, Mikoshiba K (1989) Developmental expression and intracellular location of P_{400} protein characteristic of Purkinje cells in the mouse cerebellum. Dev Biol 133:67-76

Maeda N, Niinobe M, Mikoshiba K (1990) A cerebellar Purkinje cell marker P_{400} protein is an inositol 1,4,5-trisphosphate (InsP3) receptor protein — purification and characterization of InsP3 receptor complex. EMBO J 9:61-67

Mikoshiba K, Huchet M, Changeux J-P (1979) Biochemical and immunological studies on the P_{400} protein, a protein characteristic of the Purkinje cell from mouse and rat cerebellum. Dev Neurosci 2:254-275

Mikoshiba K, Okano H, Tamura T, Ikenaka K (1991) Structure and function of myelin protein genes. Annu Rev Neurosci 14:201-217

Morell P (ed) Myelin (1984) Plenum, New York

Nakanishi S, Maeda N, Mikoshiba K (1991) Immunohistochemical localization of an inositol 1,4,5-trisphosphate receptor, P_{400}, in neural tissue: studies in developing and adult mouse brain. J Neurosci 11: 2075-2086

Supattapone S, Worley PF, Baraban JM, Snyder SH (1988) Solubilization, purification, and characterization of an inositol trisphosphate receptor. J Biol Chem 263:1530-1534

Discussion of Presentation K. Mikoshiba

Van Broeckhoven

I want to ask about the myelin-deficient mouse. Which region on the human genome does the chromosome 18 region on mouse correspond to? Secondly, do you know what kind of sequence seems to trigger this duplication formation. The reason I ask is because my group and a group in the United States recently found a duplication in Charcot-Marie-Tooth (CMT) neuropathy type I. As you know, this is a demyelinating-remyelinating neuropathy and I was very touched by the fact that you said in this myelin-deficient mouse the duplication is inverted and you can actually lose this duplication by a deletion and then have a normal gene again. In one of the families we have a normal individual, which actually could have beome a nonpatient, because this person lost the duplication. So, there are some similarities between our observation in CMT-I and the data you were presenting on this myelin-deficient mouse.

Mikoshiba

MPB gene has been assigned to the distal end of chromosome 18 in both the human and the mouse. There is unfortunately an actual disease that corresponds to human disease localized on the region. We still do not know how MPB genes are rearranged. The sequence inserted in the MPB gene is very important for the rearrangement of the gene in the immunoglobulin system, so we believe that this might also happen in the nervous system. We cannot find the sequence in the normal gene, so we presume that some kind of transposonlike sequence just entered to MBP gene and then facilitated this duplication and inversion.

Maitre

I have a question about the behavior of the mice. Are the abnormal movements you have seen sensitive to the usual drugs known to influence, for example, convulsions, ataxia or tremor or not?

Mikoshiba

We have tried to rescue the symptoms but unfortunately we could not. As far as I know, there are no drugs that rescue the symptoms.

Barnard

I also wanted to ask a question on the mld mutant. I think you have done a beautiful piece of work in identifying the duplications and inversions as an unusual mechanism for this. I would like to ask about the revertants. These are obviously somatic revertants. Was it proven that cells have in fact reverted in gene sequence? I would like to point this out as an interesting parallel to your work. In studies on Duchenne muscle dystrophy (DMD) it had been noted recently that the muscles of the patients do show a very low proportion of positive fibers for dystrophin and this has been considered a mystery. But recently, in several laboratories this has been confirmed by specific antibodies. They do seem to make dystrophin in a few cells, although dystrophin is deleted from the genome. It has been suggested that these are somatic revertants, but others have said there is no rationale for that. This is an interesting aspect as the mechanism is somewhat similar in the DMD cases; there is a large deletion and inversion is often also reported.

Mikoshiba
A similar case is also found in analbuminemia. It has been confirmed that it is an autosomal revertant, but this case is just a hypothesis. We did PCR very often. But at that region genes have accumulated and I could not get solid data to confirm that it is reverted. I think that the genes are actively moving around at that region.

Mendlewicz
I was intrigued by one of the comments you made. You said that some of the mechanisms you are describing may actually influence the diagnostic procedure using the probes. Could you elaborate on this.

Mikoshiba
If we use the probe, which is inverted, we naturally sometimes make a mistake when we do a Southern or northern blot. Sometimes genes are moving. In that case, the data do not provide definite results. And so we have to choose many probes to avoid the mistakes. That would be one solution.

Van Broeckhoven
I was referring to CMT-I because of the demyelinating-remyelinating pattern. Maybe the remyelination is due to the loss of reduplication, but from the point of human linkage studies in CMT, where there is a duplication, if we do a linkage analysis with a polymorphism which is actually within this duplication, one of the alleles shows a double density. But if you do not know that there is a duplication, you will score this individual as heterozygous and in some meioses this may lead to false recombinants. Using probes in duplications may actually lead to false results in linkage analysis studies.

Beyreuther
You mentioned that the IP_3 receptor is also altered in Alzheimer's disease. Can you explain where this receptor maps in the human chromosome and what you mean by "altered in Alzheimer disease"?

Mikoshiba
In cooperation with a group in the United States we used platelets and found that the content is altered. We are now analyzing the IP_3 locus on human chromosome.

Beyreuther
Do you see any relationship between aluminum and aluminum binding to IP_3 and this IP_3 receptor with regard to neurotoxicity of aluminum?

Mikoshiba
We have not tested it yet. That will be very interesting.

The D_3 Dopamine Receptor Gene as a Candidate Gene for Genetic Linkage Studies

P. Sokoloff, L. Lannfelt, M.P. Martres, B. Giros, M.L. Bouthenet, and J.C. Schwartz

With evidence from family, twin, and adoption studies, heredity appears to be a significant etiological factor in several neuropsychiatric disorders including schizophrenia (McGuffin 1988), bipolar affective disorder (Kidd and Weisman 1978), and Gilles de la Tourette's syndrome (Pauls and Leckerman 1986). However, the mode of inheritance of these diseases is complex and does not follow simple mendelian expectations. In addition, the probable etiological heterogeneity of diseases such as schizophrenia complicates the analysis for linkage studies. Molecular biology has provided polymorphic DNA markers, which can be classified into two types. Nonspecific approaches utilize anonymous markers along the entire genome, the number of which strongly increases as the genome mapping becomes more detailed. This approach might be fruitful, but is a long and exacting task. On the other hand, if the underlying biological substrates for these diseases can be identified, then the gene(s) governing these substrates may constitute good candidate gene(s).

Disturbances in dopamine neurotransmission and dopamine receptors have been implied in several neuropsychiatric diseases (Murphy et al. 1971; Carlsson 1988), making the dopamine receptor genes highly interesting candidates for study.

Until recently it was widely accepted that dopamine (DA) affects its target cells in brain and endocrine tissues via interaction with only two receptor subtypes, termed D_1 and D_2, each differing from the other in pharmacological specificity and in having the opposite effect on adenylate cyclase (Spano et al. 1978; Kebabian and Calne 1979). It was also generally conceded that the therapeutic efficacy of antipsychotics derives from the high affinity binding to D_2 receptors. Although we repeatedly suggested that antipsychotic agents interact to a variable extent with more than a single DA receptor subtype (Schwartz et al. 1984), this remained controversial, in spite of the substantial clinical relevance.

This situation has begun to change with the advent of molecular biology methods, which have confirmed the existence of additional DA receptors. Their existence throws a new light onto the modes of action and side effects of many drugs used in neurology and psychiatry. This is particularly the case for the D_3 receptor which we recently identified in the rat (Sokoloff et al. 1990) and human brain (Giros et al. 1990) and which appears to be a major target for antipsychotics. This paper focuses mainly on this "third receptor but data are put in perspective by comparing D_3 with the othmembers of the large DA receptor family and suggest that this receptor is a good candidate gene for linkage studies in neuropsychiatric disorders.

Molecular Biology of D_3 and Other Dopamine Receptor Genes

The first cloning of a DA receptor gene, that of the D_2 receptor (Bunzow et al. 1988), paved the way for the cloning of a series of DA receptor genes, based upon the significant sequence homology these receptors display. Quite expectedly, the D_1 receptor, which is nearly as abundant as the D_2 receptor in brain, was the first to follow (Zhou et al. 1990; Dearry et al. 1990; Sunahara et al. 1990; Monsma et al. 1990). Then came the genes of a series of less abundant and less expected receptors which markedly expand the DA receptor family: the D_3 (Sokoloff et al. 1990), D_4 (Van Tol et al. 1991), and D_5 receptors (Sunahara et al. 1991); the main features of these receptors are summarized in Table 1. The amino acid sequence of all these receptors, as deduced from their established nucleotide sequence, reveals that they belong to a larger superfamily, that of receptors with seven transmembrane domains (TMs), and are coupled to their intracellular transduction system by a G protein.

Table 1. Synopsis of dopamine receptor subtypes

	D_1	D_2	D_3	D_4	D_5
Coding sequence	446 a.a.	D_{2A} = 443 a.a. D_{2B} = 414 a.a.	400 a.a.	387 a.a.	477 a.a.
Chromosome localization	5 q31-q34	11 q22-q23	3 q13.3	11 p	4 p
Highest brain densities	Neostriatum	Neostriatum	Paleostriatum (islands of Calleja, nucleus accumbens)	Medulla Frontal cortex	Neostriatum Hippocampus
Pituitary gland	No	Yes	No	?	No
Dopamine neurons (A9, A10)	No	Yes	Yes	?	?
Affinity for dopamine	Micromolar	Micromolar	Nanomolar	Submicromolar	Submicromolar
Characteristic antagonist	SCH-23390	Haloperidol	UH 232	Clozapine	SCH-23390
Adenylyl cyclase	Stimulates	Inhibits	?	?	Stimulates
Characteristic agonist	SKF-82526	Bromocriptine	Quinpirole	?	SKF-82526

The various genes of the DA receptor family can be classified into two groups according to their organization: (1) intronless genes, i.e., those of the D_1 and D_5 receptors, in which the coding nucleotide sequence is continuous; and (2) genes whose coding sequence is contained in discontinuous DNA segments (exons) interspersed among sequences (introns) that do not form a part of the mature mRNA. This last organization, found in the rhodopsin gene or in the D_2, D_3 and D_4 receptor genes (Fig.1), may potentially lead to the biosynthesis of several distinct proteins encoded by a unique gene (see below) via a mechanism of alternative splicing (in which a given exon in the pro-mRNA is either present or absent in the final mRNA).

The molecular cloning of the D_3 receptor involved a combination of screenings of cDNA and genomic libraries and reverse transcription-polymerase chain reaction (PCR; Sokoloff et al. 1990). The open reading frame of the D_3 receptor corresponds to a sequence of 446 amino acid residues in the rat but only 400 residues in humans (Giros et al. 1990), the main difference residing at the level of the third putative intracytoplasmic loop (i_3). There is relatively little sequence homology at the level of this loop between D_2 and D_3 receptors, contrasting with the high amino acid sequence homology at the level of the TMs where the dopaminergic ligands are throught to bind: for instance, homology at this level is as high as 78% between the human D_2 and D_3 receptors.

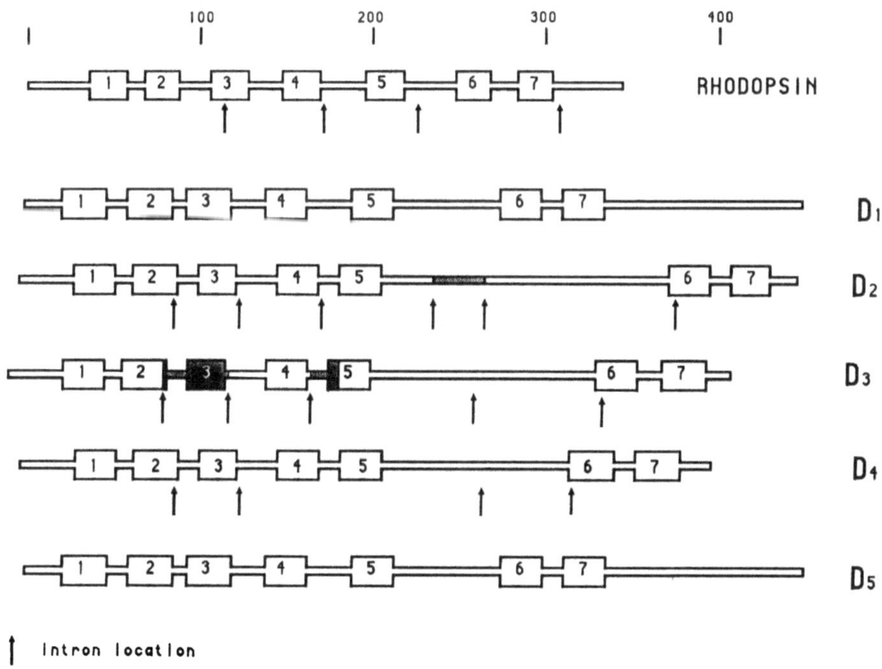

Fig.1. The family of dopamine receptor genes: organization and comparison with rhodopsin. The *scale* indicates the length of the amino acid sequence. *Arrows* indicate the position of introns; *shaded areas* correspond to alternative exons

The structural knowledge derived from physical studies of the opsins (Findlay and Eliopoulous 1990) and discrete manipulations of the ß-adrenergic receptor gene ("site-directed mutagenesis") (Strader et al. 1989) can be extended to other members of the superfamily, e.g., the DA receptors, particularly the human D_3 receptor. They all comprise a pattern of seven stretches of 20 to 25 hydrophobic amino acids postulated to form transmembrane α-helices, connected by alternating extracellular and cytoplasmic loops constituted by hydrophilic residues. The N-terminal part constitutes a glycosylated extracellular domain. The transmembrane helices constitute the ligand binding domain, particularly three amino acid residues thought to interact with catecholamines: an aspartic residue (Asp^{110} in the human D_3 receptor) in TM3, which forms an ion pair with the protonated amine group of DA; and two serine residues (Ser^{193} and Ser^{196}) in TM5, which presumably form a hydrogen bonding interaction with the two phenol groups of DA. This last interaction, specific for DA and agonists, could cause a conformational change in the helix, transmitted to the i_3 loop involved in the coupling to the G protein, and thereby transduce the signal.

Splice Variants of D_3 Receptor mRNA

Alternative splicing was shown to occur in the case of the D_2 receptor, potentially leading to two distinct receptors differing by a stretch of 29 amino acids at the level of i_3, which were called $D_{2(444)}$ (or D_{2L} for D_2 long and D_{2A}) and $D_{2(415)}$ (or D_{2S}, for D_2 short and D_{2B}). These two isoforms of the D_2 receptor are identical pharmacologically but are expressed differently in the different cerebral areas, may interact differently with various G proteins (Giros et al. 1989; Dal Toso et al. 1989), and their relative abundance is affected by neuroleptic treatments (Martres et al. 1991).

In the case of the D_3 receptor, alternative splicing gives rise potentially, in addition to the 446 amino acid receptor, to two truncated proteins of 109 and 428 amino acids (Giros et al. 1991). Thus, PCR amplification, using primers flanking the entire coding sequence of the D_3 receptor gave rise in various rat brain areas to two other products with sizable deletions of 113 bp in TM3 and 54 bp in 02, respectively: hence the designation of the proteins potentially encoded by these two transcripts as D_3 (TM3-del) and D_3(02-del), respectively (Fig. 1).

Two distinct alternative splicing mechanisms underlie the production of these two mRNAs. In the case of D_3 (TM3-del), the process involves combinatorial exons, the "cassette" exon being the second exon (Fig. 1). Since the latter does not comprise nx3 nucleotides, this introduces a frame shift in the sequence and the splice product encodes a 109 amino acid protein. By contrast, in D_3(02-del) mRNA, the in-frame 54 bp deletion does not correspond to a full exon: alternative splicing occurs within the fourth exon where an internal acceptor site can be used by the splicing machinery, thereby giving rise to an mRNA encoding a 428 amino acid protein.

Whereas the structure of D_3(TM3-del) makes it unlikely that the protein may function as a receptor, this is not so clear in the case of D_3(02-del), whose structure may still be compatible with the occurrence of seven TMs, as revealed by the hydropathy profile. However, CHO clones stably expressing D_3(02-del) mRNA failed to show any dopaminergic binding activity, as assessed with various radioactive ligands.

What could be the function, if any, of these truncated products of the D_3 receptor gene? Indeed, both encode potential integral membrane proteins possibly involved in cell signaling. Nevertheless the idea that these truncated forms lack any direct biological activity in signal transduction cannot be discarded. They could be formed at random during biosynthesis of the functionally active D_3 receptor. Alternatively, this may represent a mechanism controlling the abundance of the active D_3 receptor.

Finally, since multiple D_3 receptor gene transcripts are also found in human brain (Giros et al. 1990), it cannot be excluded that defects in the alternative splicing mechanisms, leading to the formation of inactive receptors, might occur during psychiatric diseases.

Anatomical Distribution of D_3 Receptor mRNA in Rat Brain

The distribution of D_3 receptor gene transcripts in rat brain areas, as established using northern or PCR analysis or visualized by in situ hybridization histochemistry (Sokoloff et al. 1990; Bouthenet et al. 1991), markedly differs from those of the D_1 (Fremeau et al. 1991) or D_2 receptor (Meador-Woodruff et al. 1989) gene transcripts. For instance, only a weak D_3 receptor hybridization signal was detected in restricted parts of the striatum, while the whole striatum contains the highest densities of DA axons and D_2 receptor mRNA (Fig. 2). By contrast, the D_3 receptor mRNA is highly

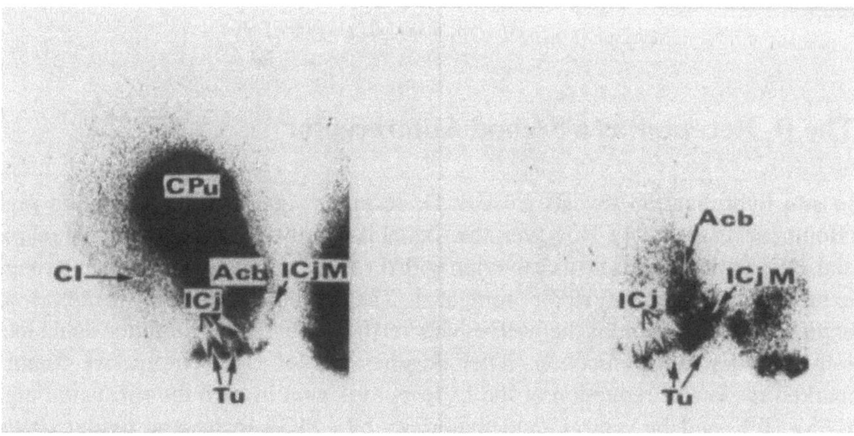

Fig. 2. Compared distributions of D_2 (*left*) and D_3 (*right*) receptor mRNAs established by in situ hybridization in frontal sections performed in rat telencephalon. Note the nonoverlapping complementary distributions of the two transcripts in the ventral striatum, particularly at the level of olfactory tubercle, islands of Calleja. *Acb*, accumbens nucleus; *ICj*, islands of Calleja; *ICjM*, island of Callera major; *CPu*, caudate putamen; *Tu*, olfactory tubercle

expressed in the olfactory tubercle, island of Calleja complex, and the nucleus accumbens. These areas constitute, with the ventral and ventromedial parts of the caudate putamen, the "ventral striatum," a territory receiving afferents from the prefrontal or allocortex and amygdala and its major DA inputs from the A10 cell group in the ventral tegmental area. It projects to ventral pallidum and the latter to the mediodorsal thalamic nucleus which selectively innervates the prefrontal cortex (Björklund and Lindvall 1984). This connectivity has led to the designation of this territory as the "limbic" part of the striatal complex, in which D_3 receptors may, therefore, mediate a large part of DA signals. The remainder of the striatal complex, which is mainly innervated by DA projections from the substantia nigra, receives its cortical inputs from the somatic neocortex and is highly enriched in D_1 or D_2 receptors. D_3 receptor signals were also detected in other "limbic" areas such as the hippocampus, septum, or mammillary nuclei in the hypothalamus. This suggests a major participation of D_3 receptors in dopaminergic transmissions in limbic areas known to be associated with cognitive, emotional, and endocrine functions.

D_2 receptor mRNA is also highly expressed in these areas but there is *no strict overlap* with D_3 receptor mRNA distribution: for instance, the highest levels of D_3 receptor mRNA in brain are detected in the islands of Calleja, in which the D_2 receptor signal is weak, whereas a reverse situation is found in the olfactory tubercles (Fig. 2). The two receptor subtypes differ by the much higher affinity of dopamine for the D_3 receptor and, possibly, by their intracellular signaling systems (see below). Hence, it seems likely that different kinds of signals might be generated by DA in neighboring but topographically distinct cerebral structures.

Interestingly, no specific D_3 receptor signal could be detected by northern and PCR analyses in the pituitary gland, a prototypical localization of D_2 receptors. This suggests that selective D_3 receptor ligands, when available in therapeutics, will not affect the activity of mammotrophs as do the neuroleptics which are currently being used.

The D_3 Receptor as a Second Autoreceptor

In situ hybridization reveals a weak D_3 receptor signal in the substantia nigra (Bouthenet et al. 1991). However, this signal is mainly expressed in the lateral part and, here again, there is no true overlap with the D_2 receptor signal, which is strongly expressed all over the whole compacta. The hypothesis that D_3 receptors are expressed by DA neurons themselves was verified after their destruction using local 6-hydroxydopamine injection. After degeneration of DA neurons, we found a marked ipsilateral reduction of the D_3 receptor signal in both the substantia nigra ($-65 \pm 10\%$) and the ventral tegmental area ($-69 \pm 14\%$). In the same tissue extracts, the D_2 receptor mRNA levels were similarly affected, i.e., by - 88% and - 65%, respectively (Sokoloff et al. 1990).

This establishes not only that both D_2 and D_3 receptors are located postsynaptically but also that they are expressed by DA neurons belonging to the A_9 and A_{10} cell groups. This suggests that both have a role as autoreceptor. Such a role for the D_3 receptor is consistent with its pharmacological profile (see below).

Many distinct functions were previously attributed to DA autoreceptors, i.e., inhibitions of impulse flow, DA synthesis and release at either nerve terminals or dendrites, and cotransmitter release. D_2 and D_3 autoreceptors might in various ways participate in all these actions and in various brain areas. Finally, the question as to whether a single cell expresses both D_2 and D_3 receptors remains to be answered, namely, by in situ hybridization studies at the cellular level.

Pharmacology of the D_3 Receptor

The pharmacology of the rat (Sokoloff et al. 1990) or human D_3 receptor (P. Sokoloff, M. Andrieux, R. Besançon, C. Pilon, M.P. Martres, B. Giros, and J.C. Schwartz, unpublished data, 1991) was studied in transfected CHO cells expressing a high level of sites labeled with high affinity by [^{125}I]iodosulpride, formerly considered to be a D_2 receptor-selective ligand (Martres et al. 1985).

The D_3 receptor can be considered, like the D_4 receptor, to be a "D_2-like" receptor: it poorly recognizes "D_1-specific" ligands such as SKF 38393 or SCH-23390, whereas it binds "D_2-specific" agonists, e.g., quinpirole, or antagonists, e.g., sulpiride (Table 2). However, several salient pharmacological features of the D_3 receptor should be underlined.

1. Dopamine as well as agonists such as TL99, quinpirole, or quinerolane display high affinities at D_3 receptors. This may account for the role of D_3 receptors as autoreceptors since DA in very low concentrations reduces the activity of DA neurons, and these agonists seem to act preferentially at presynaptic autoreceptors as judged in animal models, such as the butyrolactone-induced increase of DA synthesis (Wolf and Roth 1987). This suggests that some functions attributed to autoreceptor stimulation actually involve the D_3 receptor. In agreement, AJ76 and UH232, the only antagonists exhibiting (limited) D_3 receptor selectivity, have behavior-stimulating properties in animals, attributed to autoreceptor blockade (Svensson et al. 1986). These pharmacological data suggest that the D_3 receptor plays a major role in the feedback inhibition of DA transmission.

2. Most antipsychotics display high affinities at the D_3 receptor, indicating that this receptor is probably blocked during the treatment of schizophrenia and related disorders. The degree of this blockade would depend, however, on the antipsychotics used, since their recognition by the D_3 receptor relative to that of the D_2 receptor is variable. The compounds for which the ratios between K_i values for D_2 and D_3 receptors (K_iD_2:K_iD_3 ratios) are the highest would exert a more complete blockade of DA transmission in limbic areas, where the D_3 receptor is selectively expressed. Conversely, those for which the ratios are the lowest would preferentially block the D_2 receptor present in other dopaminergic areas, including the extrapyramidal system, mainly involved in the control of motor function. This could be one of the molecular bases for distinguishing "atypical" neuroleptics. Consistent with this hypothesis is the observation of a high K_iD_2:K_iD_3 ratio measured with atypical

Table 2. Pharmacology of dopamine receptor subtypes

	K_i Values (n*M*)				
	D_1	D_2	D_3	D_4	D_5
Agonists					
Dopamine[a]	2300	2000	30	450	230
Apomorphine	680	70	70	(4)[b]	360
Bromocriptine	700	5	7	500	500
Pergolide	1400	20	2	-	900
Quinpirole	> 20000	1400	40	50	> 20000
SKF 38393	150	10000	5000	10000	100
Antagonists					
Haloperidol	30	0.6	3	5	40
Pimozide	-	10	11	40	-
(-)Sulpiride	40000	10	20	50	80000
UH 232	-	40	10	-	-
Clozapine	140	70	500	9	250
SCH-23390	0.3	1000	1000	3500	0.3

Values for human D_2 and D_3 receptors transfected in CHO cells are from this laboratory whereas those for D_1, D_4 and D_5 receptors were taken from Van Tol et al. (1991) and Sunahara et al. (1991).
[a]Value + Gpp(NH)p.
[b]High affinity component.

neuroleptics such as sulpiride or amisulpiride. Nevertheless the peculiar clinical properties of clozapine are more likely to derive from its higher affinity for D_4 than any other receptor subtype (Van Tol et al. 1991).

Interestingly, among antipsychotics having the highest K_iD_2:K_iD_3 ratios are amisulpiride, carpipramine, pipothiazine, and pimozide, which all exhibit definite desinhibitory actions sought in the treatment of negative symptoms in schizophrenia. Conceivably, the more efficient blockade of D_3 autoreceptors by these compounds could facilitate DA transmission in some brain areas, which might lead to the alleviation of negative symptoms (Carlsson 1988). To address these questions, further studies will be necessary, however, using more selective compounds, the design of which should be facilitated by the use of clonal cell lines expressing a single receptor subtype.

Genetic Studies with the D_3 Dopamine Receptor

The aforementioned properties of the D_3 dopamine receptor gene suggest that it plays a crucial role in the neurotransmission of DA in the limbic brain areas, where disturbances of dopaminergic systems have been proposed to participate in the etiology of several neuropsychiatric disorders. Although the involvement of D_3 receptor dysfunctions in these disorders remains to be established, the use of drugs acting at this receptor significantly contributes to the management of mental illnesses.

In addition, the complexity of the D_3 receptor gene and its regulation, possibly involving alternative splicing of the pro-mRNA, raises the possibilities of inherited defects in its function. Hence, the D_3 receptor gene should be regarded as a good candidate gene for genetic linkage studies in neuropsychiatric diseases. To explore this hypothesis, we have mapped the D_3 receptor gene and started to identify polymorphisms which can be used as gene markers.

The human D_3 receptor gene was localized to chromosome 3q13.3 by in situ hybridization (Le Coniat et al. 1991). Several restriction fragment length polymorphisms (RFLP) were identified on the gene, among which a two-allele Bal I polymorphism was detected by direct sequencing of the alleles (Lannfelt et al. 1991). This mutation substitutes an amino acid residue in the N-terminal extracellular domain. The Bal I polymorphism was studied by PCR and found to be inherited in a codominant mendelian mode. The high frequencies of homozygotes in a normal population (41% and 15% for the two alleles, respectively, compared to 44% heterozygotes) suggest that the mutation has no deleterious consequences, a hypothesis supported by the localization of this mutation outside the functional domains of the receptor.

Hence, the Bal I polymorphism indeed constitutes a useful genetic marker of the D_3 receptor gene and of loci in its vicinity along the third chromosome. It is currently beeing used in several laboratories in linkage studies with different pathologies.

Conclusions

The existence of the three pharmacologically distinct "D_2-like" subtypes, i.e., D_2, D_3 and D_4, instead of a single D_2 receptor, which was formerly recognized as *the* target for antipsychotic agents, raises important issues: of the three, which is (are) responsible for the beneficial therapeutic effects? Which is (are) responsible for each unwanted side effect? At this early stage of our knowledge much caution is needed but two clues point to D_3 as a key receptor in schizophrenia and other neuropsychiatric disorders: its selective expression in a phylogenetically old part of the brain known as the limbic system and its relatively preferential binding of several atypical antipsychotics. The probable role of the D_3 receptor as a main target for antipsychotics raises the possibility that its corresponding gene might be affected in various psychiatric diseases, a hypothesis being actively explored at present in several laboratories.

References

Björklund A, Lindvall O (1984) Dopamine-containing systems in the CNS. In: Björklund A, Hökfelt T (eds) Classical transmitters in the CNS. Elsevier, Amsterdam, pp 55-122 (Handbook of chemical neuroanatomy, vol 2)

Bouthenet ML, Souil E, Martres MP, Sokoloff P, Giros B, Schwartz JC (1991) Localization of dopamine D_3 receptor RNA in the rat brain using in situ hybridization histochemistry: comparison with dopamine D_2 receptor mRNA. Brain Res (in press)

Bunzow JR, Van Tol HHM, Grandy DK, Albert P, Salon J, Christie McD, Machida CA, Neve KA, Civelli O (1988) Cloning and expression of a rat D_2 dopamine receptor cDNA. Nature 336:783-787

Carlsson A (1988) The current status of the dopamine hypothesis of schizophrenia. Neuropsychopharmacology 1:179-186

Dal Toso R, Sommer B, Ewert M, Herb A, Pritchell DB, Bach A, Shivers BD, Seeburg PH (1989) The dopamine D_2 receptor: two molecular forms generated by alternative splicing. EMBO J 8:4025-4034

Dearry A, Gingrich JA, Falardeau P, Fremeau RT, Bates MD, Caron MG (1990) Molecular cloning and expression of the gene for a human D_1 dopamine receptor. Nature 347:72-76

Findlay J, Eliopoulous E (1990) Three-dimensional modelling of G protein-linked receptors. Trends Pharmacol Sci 11:492-499

Fremeau RT, Duncan GE, Fornaretto MG, Dearry A, Gingrich JA, Brees GR, Caron MG (1991) Localization of D1 dopamine receptor mRNA in brain supports a role in cognitive, affective, and neuroendocrine aspects of dopaminergic neurotransmission. Proc Natl Acad Sci USA 88:3772-3776

Giros B, Sokoloff P, Martres MP, Riou JF, Emorine LJ, Schwartz JC (1989) Alternative splicing directs the expression of two D_2 dopamine receptor isoforms. Nature 342:923-926

Giros B, Martres MP, Sokoloff P, Schwartz JC (1990) cDNA cloning of the human dopaminergic D3 receptor and chromosome identification. CR Acad Sci (Paris) III:501-508

Giros B, Martres MP, Pilon C, Sokoloff P and Schwartz JC (1991) Shorter variants of the D_3 dopamine receptor produced through various patterns of alternative splicing. Biochem Biophys Res Commun 176:1584-1592

Kebabian JW, Calne DB (1979) Multiple receptors for dopamine. Nature 277:93-96

Kidd KK, Weisman MM (1978) Why do we not understand the genetics of affective disorders. In: Cole JJ, Schurtzberg AF, Frazier SH (eds) Depression: biology, psychodynamics and treatment. Plenum, New York, pp 107-121

Lannfelt L, Sokoloff P, Martres MP, Pilon C, Giros B and Schwartz JC (1991) The dopamine D_3 receptor gene and schizophrenia. Psychiatr Genet 2:16

Le Coniat M, Sokoloff P, Hillion J, Martres MP, Giros B, Pilon P, Schwartz JC, Berger R (1991) Chromosomal localization of the human D_3 dopamine receptor gene. J Am Hum Genet (in press)

Martres MP, Bouthenet ML, Salès N, Sokoloff P and Schwartz JC (1985) Widespread distribution of brain dopamine receptors evidenced with ^{125}I-iodosulpride, a highly selective ligand. Science 228:752-755

Martres MP, Sokoloff P, Giros B, Schwartz JC (1991) Effects of dopaminergic transmission interruption on the D_2 receptor isoforms in various cerebral tissues. J Neurochem (in press)

McGuffin P (1988) Genetics of schizophrenia. In: Beddington P, McGuffin P (eds) Schizophrenia: the major issues. Heinemann Medical, London, pp 107-126

Meador-Woodruff JH, Mansour A, Bunzow JR, Van Tol HHM, Watson SJ, Civelli O (1989) Distribution of D_2 dopamine receptor mRNA in rat brain. Proc Natl Acad Sci USA 867625-7628

Monsma FJ, Mahan LC, McVittie LD, Gerfen CR, Sibley DR (1990) Molecular cloning and expression of a D_1 dopamine receptor linked to adenylyl cyclase activation. Proc Natl Acad Sci USA 87:6723-6727

Murphy DL, Brodie HKH, Goodwin FK (1971) Regular indication of hypomania by L-dopa in "bipolar" manic-depression patients. Nature 229:135-136

Pauls DL, Leckerman JF (1986) The inheritance of Gilles de la Tourette syndrome and associated behavior. N Engl J Med 316:993-997

Schwartz JC, Delandre M, Martres MP, Sokoloff P, Protais P, Vasse M, Costentin J, Laibe P, Wermuth CG, Gulat C, Lafitte A (1984) Biochemical and behavioral identification of discriminant benzamide derivatives: new tools to differentiate subclasses of dopamine receptors. In: Usdin E, Carlsson A, Dahlstrom A, Engel J (eds) Catecholamines: neuropharmacology and central nervous system. Theoretical aspects. Liss, New York, pp 59-72 (Neurology and neurobiology, vol 8B)

Sokoloff P, Giros B, Martres MP, Bouthenet ML, Schwartz JC (1990) Molecular cloning and characterization of a novel dopamine receptor (D_3) as a target for neuroleptics. Nature 347:146-151

Spano PF, Govoni S, Trabucchi M (1978) Studies on the pharmacological properties of dopamine receptors in various areas of the central nervous system. Adv Biochem Psychopharmacol 19:155-165

Strader DC, Sigal SI, Dixon AFR (1989) Mapping of functional domains of the β-adrenergic receptor. Am J Respir Cell Mol Biol 1:81-86

Sunahara RK, Niznik HB, Weiner DM, Stormann TM, Brann MR, Kennedy JL, Gelernter JE, Rozmahel R, Yang Y, Israel Y, Seeman P, O'Dowd BF (1990) Human dopamine D_1 receptor encoded by an intronless gene on chromosome 5. Nature 347:80-83

Sunahara RK, Guan HC, O'Dowd BF, Seeman P, Laurier LG, Ng G, George SR, Torchia J, Van Tol HHM, Niznik HB (1991) Cloning of the gene for a human dopamine D_5 receptor with higher affinity for dopamine than D_1. Nature 350:614-619

Svensson K, Johansson AM, Magnusson T, Carlsson A (1986) (+)-AJ 76 and (+)-UH 232: central stimulants acting as preferential dopamine autoreceptor antagonists. Naunyn Schmiedebergs Arch Pharmacol 334:234-245

Van Tol HHM, Bunzow JR, Guan HC, Sunahara RK, Seeman P, Niznik HB, Civelli O (1991) Cloning of the gene for a human dopamine D_4 receptor with high affinity for the antipsychotic clozapine. Nature 350:610-614

Wolf ME, Roth RH (1987) Dopamine autoreceptors. In: Creese I, Fraser CM (eds) Dopamine receptors, vol 8. Liss, New York, pp 45-96

Zhou QZ, Grandy DK, Thambi L, Kushner JA, Van Tol HHM, Cone R, Pribnow D, Salon J, Bunzow JR, Civelli O (1990) Cloning and expression of human and rat D_1 dopamine receptors. Nature 347:76-80

Discussion of Presentation P. Sokoloff et al.

Ackenheil
After these two presentations it is evident that the kind of mutation is very important for the function. Point mutation may not have any functional effect; on the other hand it can have a substantial effect. Relating these findings to clinical diagnoses and to linkage studies, this may become a serious problem because the function might not be changed due to a point mutation. Would it not be better to see whether there is a functional defect and then do linkage studies?

Sokoloff
I agree that a mutation or a defect in other genes related to the dopamine receptor may affect the transmission. But it is a problem in all linkage studies that there may be an interaction between genes. There are, however, many technical problems because the frequency of naturally occurring mutations can be high. And in some cases it is very hard to determine the exact position in the sequence. In the case of the D_3 we detected by chance one point mutation with in the coding sequence that is in a core position. But in other cases, I think it is very difficult to determine the position of the mutation and its possible functional consequence prior to performing linkage studies.

Caron
Can I make a comment on that. Even when you say that this mutation is on the region of the receptor whose function we do not really understand very well, I would like to remind you that in the rhodopsin gene Elliot Burson at Harvard has characterized mutants that are almost at the same region as this receptor and the mutation that you showed. This is presumably the molecular basis of some form of retinitis pigmentosa, which leads to degeneration of the retina. So this region of the receptor might be important in the processsing. Before we examine all of the potential roles of these things, we just should keep our optimism and keep going.

Sokoloff
The mutation is found at a high frequency in the normal population.

Kidd
I have two comments. It is well known in human genetics that there are many examples of allelic heterogeneity, for there are alleles that are subnormal but not necessarily leading to any clinical malformation by themselves. But possibly in a

heterozygote with the more severely abnormal allele will it greatly modify the expression, so that you can get a lot of variation in expression. Well known are the hyperphenylalaninemia mutations at the phenylalanine hydroxylase locus, where you get children who are not quite classified as having PKU but clearly have abnormal phenylalanine hydroxylase levels. The other comment is on the publication on Tourette syndrome. We published a brief note, pointing out that the analysis did not use the data quite properly and in fact assumed an incorrect linkage map in reaching the high lod score, so that only some of the pairwise lod scores are really valid. They do not reach statistical significance and there are several other laboratories that have quite strongly negative data in that region.

Reich

Examining the "attempting to fit models" form to segregation data for either manic depression or schizophrenia, one can easily rule out a single major locus model. However, one begins to think about two- or three-way epistatic models. The genetics implies at least two or more defects in the same functional system to result in a disorder. Now, you have shown us complex systems, which surely could have more than one defect. I wonder whether or not there are some complex, perhaps dynamic, studies of these systems and how they work. Perhaps not one lesion would do, but two or three.

Sokoloff

You have to consider that this situation may be more simple. In fact, the various defects can lead to only two situations: an increased or decreased function.

Crow

I would like to ask a question about the genetics of both speakers. Do we have sufficient information across species, including primates other than humans, as to whether any of these receptors have been evolving fast recently.

Sokoloff

It seems that the receptors containing introns evolved more tardively.

Caron

In my case I do not think we know. There are several species that are now being cloned for betas and alphas. But I do not hink there is enough information on that.

Maybe I can give you one example. We have recently found that in the D_1 receptor family there appear to be pseudogenes, and I guess they are "pseudo" because they have stop codons in their coding sequence. But what is amazing to us is that the sequence on either side of the codon is almost 100% conserved. So they probably do not evolve very fast. Others have characterized one of the pseudogenes and there are several examples, I guess, in the D_1 receptor family. They at least go back to 8 million years ago and it was not particularly different. So they do not evolve that fast.

Vogel
Coming back to your statement that this D_3 mutant is relatively common in the population. Did you look into the problem of whether this might have an influence on normal psychology, because we are not only interested in the disease but also in genetic variation in the normal range. If you do think this has an influence on brain function, such results should induce somebody to look into normal variation within the psychological range.

Sokoloff
I agree with you, but now it is your work to find out.

Wildenauer
You said your second splice mutant might be involved in regulation. Have you looked at the tissue distribution of this or have you compared it with the original?

Sokoloff
We have examined the shorter isoforms in several tissues but we did not find variation between brain tissues as in the case of D_2 receptor isoforms.

Barnard
I just want to make a comment and reply to Tim Crow's question. I think that at least for those receptors which are very widely used, such as the GABA receptor, the evidence is that they are evolving extremely slowly. For with the GABA receptors, we have now recently cloned DNAs through several marmots and birds, and the rate of change is extremely low. With RFLPs we further looked for cDNAs through a large panel of humans and found none at all by the most obvious criteria that everybody has the same GABA receptors in the brain. The amount of variation in the population is very small.

Racagni
Do you have any ideas on the coupling of the D_3 receptor to a G_i or another? And secondly, some antipsychotics, such as pimozide and others, have a high affinity for the D_3 receptor and these are compounds which many use in treating the negative symptoms of schizophrenia. Thus, the D_3 receptor might be more involved in this pathology than in the other.

Sokoloff

We expressed the D_3 receptor in the CHO cell line but did not observe any effects on adenylate cyclase or phospholipase C or A_2. This cell line contains high levels of G_i protein, the three subtypes of known G_i proteins, and so the D_3 receptor probably does not act through a G_i protein. This CHO cell line is a very good model to search for functional coupling, because we can introduce in it several G protein genes. Thus we are currently trying to determine a coupling mechanisms using the CHO cell line transfected with the a subunit of G_0. But we do not have data as yet.

Secondly, as you have observed, pimozide and other compounds that are used in the treatment of the negative symptoms have the highest ratio between K_i value for D_2 and D_3 receptors and this may have important implications. This also supports the role of the D_3 receptor as autoreceptor, because if you block a receptor whose normal function is to reduce the dopaminergic activity, then you activate dopamine neurotransmission. Perhaps this is what is to look for in these negative states.

Mendlewicz

From the clinical point of view, I do not think that we have clear evidence that pimozide or any other neuroleptics are superior in the treatment of the negative symptoms of schizophrenia at present. What we can accept is that some neuroleptics are better tolerated, have fewer side effects, are less sedating, cause fewer extra-pyramidal symptoms, but there is no definitive evidence today that they are superior in the treatment of negative symptoms. On the other hand, clozapine, which seems to be more promising in treating this subgroup of schizophrenia, has a low affinity for D_3 receptor.

Sokoloff

I agree that this situation is not so clear. But, pimozide or pipothiazine is currently being used in the treatment of these symptoms, at least by some psychiatrists.

Discussion of Presentation M. G. Caron: Structure and Function Relationships of the G-Receptor Family

Marc Caron presented an overview of structure and function relationships of the G-receptor family, mechanisms of coupling to the second messenger system, and desensitization of these receptors. He further reported about an observation which suggests that point mutations in these receptors might contribute to changing their function and might thus be highly interesting for CNS disorders.

The Editors

Barnard

Can you interpret the structure of requirements that you find for this loss of desensitization. They seem rather strange. For example, when you substituted to get a glutamate, you had very high activity. But when you put an aspartate in, it was almost the same as the wild type. Is there any model for what is happening there?

Caron

If you look at the various properties of the amino acid that we have put back into this position, there is really no correlation with any of the properties. What we have done in mutating this site is to destroy essentially a secondary or tertiary structure which is important in keeping the receptor there. But if you look at charge residues versus hydrophobicity, hydrophilicity, there is really no correlation that you can draw from that. The only other example that we know of where this has happened, and this has been examined, is in the RAS protein. If you substitute all 19 amino acids, 18 of them are activated, and essentially there is one proline which does the work in that system. So this is very similar to that.

Reich

One of the possible consequences of the intense homology between the various structures would be a relatively nonspecific relationship between some of the enzymes, a ß-adrenoreceptor kinase, for example, and perhaps a rhodopsin kinase. I was a bit worried that maybe that means there would never be very specific kinds of treatment, very specific kinds of intervention, because of the relatively nonspecific nature of the structure itself.

Caron

That is a good question, because we know that these molecules probably also represent families. We have cloned at least two members of the Bark family and cloned two members of the Borestan family. But, how far they extend we really do not know. We are now trying to sort out the specificity between these few kinases, which we have called bark 1 and bark 2, for receptors that have a long C-terminal tail, as in the example of the ß-receptors. We know that other receptors of the type which have a long third cytoplasmic loop and almost no C-terminal tail may also be substrates for these kinases. So there are possibly two kinases for that. But as far as having kinases that are absolutely specific for a given receptor, I do not think that is going to happen. There is really no activation of these kinases; we think that there is a substrate activation. The receptor becomes a substrate when it is occupied and that controls the activity of this enzyme. Or maybe there is some recruitment of the enzyme to the membrane. So you could say that this may serve essentially as the triggering mechanism and there may not be a kinase for each receptor. As to whether one could not intervene with that, I do not agree with you. In a system where the desensitization was deranged, you will affect the system that is deranged before the normal ones. Because in normal systems there are a lot of other steps that have been built in as safety steps.

Maitre

Concerning the desensitization process and the example you have shown with isoproterenol, is it to be understood that practically all levels of desensitization could be present in a normally functioning brain, depending on whether the receptors are located relatively near to or far from the agonist-releasing synapses?

Caron

I think we are just beginning to get answers to that question. Obviously now with probes for the receptor kinases we can address some of these questions. And recently we and Sol Snyder started doing cytochemistry of these kinases in our labs. The preliminary results seem to indicate that they localize at synapses. These kinases are mostly found in innervated tissue and in brain. So the hypothesis that these are more involved at synaptic transmission is becoming more and more reasonable. But, it is not to say that signal transduction at nonsynaptic mechanisms is not regulated. We think that cAMP is more involved in the regulation of peripheral receptors, whereas bar mechanism may be more involved at synaptic receptors. This is based on the fact that if you only occupy 1% or 2% of the receptors, as you would have in the periphery because of the low concentration of circulating catecholamines, you are not going to make them substrate. You are thus not going to phosphorylate 100% of the receptor. However, you raise cAMP and activate cAMP-dependent kinase to its maximum and then they beome substrate, whereas when you release a bolus amount of transmitter in the brain, you do not activate 100% of receptors and bark should be the most important mechanism at that time. We have evidence that bark mechanisms are more rapid in cAMP-dependent mechanism. Those things are difficult to prove.

Mallet

In considering the diversity of these receptors and with respect to the tremendous capacity for adaptation in the nervous system, have you tried to knock some of them down and to see the effect in mice.

Caron

No, we have not. We are just getting into the process of doing transgenic mice.

Racagni

What you have presented is also very important to consider for the functional state of the receptors. This depends also on the affinity for the agonist, because not all the receptors have the same affinity to all the same agonists. In a certain area you may have a situation in which you may have some receptors which are desensitized and therefore may be inactive and other receptors which may be active, depending on the affinity to the neurotransmitters in that area. Therefore it is also important to establish the receptor–receptor interaction, because through other receptors you may phosphorylate a receptor and therefore you can inactivate it. So it becomes very difficult to study the functional state of receptor in a certain time.

Caron

I could not agree more with you. I think it is easy to make hypotheses and very hard to prove them.

Racagni

And to transfer these data into the linkage study in order to have specific probes would be very difficult.

Caron

To have the probes and to do the linkage study is very easy, but what is not so easy is to convince oneself essentially that this is worth doing. We can develop a lot of theories, but it is very diffcult to show in living organisms what is more important.

GABA Receptor Genes: Applications to Neuropsychiatric Disorders

E.A. Barnard

The $GABA_A$ Receptors

The major inhibitory neurotransmitter in the vertebrate brain, γ-aminobutyric acid (GABA), is released onto the order of 70% of neurons there. Fast inhibitory neurotransmission is mediated by GABA by its binding to the hetero-oligomeric $GABA_A$ receptor and opening there an integral chloride-ion channel (Barnard et al. 1987). Molecular biology studies have revealed the existence of at least four different subunit types for this receptor (designated α, β, γ and δ) and, further, a series of isoforms for three of these subunits (α1 to α7, β1 to β4, and γ1 to γ3). These can assemble to form a range of receptor subtypes (reviewed by Olsen and Tobin 1990; further references are cited by Glencorse et al. 1991a, b). More specific isoforms may yet be discovered, as exemplified by ρ1 and ρ2, recently characterised as retinal GABA receptor subunits (Cutting et al. 1991). Various combinations of α plus β, or of α plus β plus γ subunit isoforms can form functional receptors showing a range of pharmacological properties, as is known by expressing them in recombinant form in non-neuronal cells (reviewed by Sigel et al. 1990). In particular, the γ2 subunit confers on the receptor the property of strong potentiation of the GABA response by benzodiazepine positive agonists and inhibition of it by benzodiazepine "inverse agonist" drugs (Luddens and Wisden 1991). However, the actual subunit composition of an in vivo $GABA_A$ receptor subtype is unknown.

In situ hybridisation histochemistry of the subunit mRNAs is at present the most feasible approach to investigate the forms of this receptor which occur at different locations in the brain (Wisden et al. 1988; Luddens and Wisden 1991). No unique and general patterns of colocalisation have been found, with different brain regions varying in the mixture of isoforms present, and with different neurons in one population sometimes differing in their repertoire of these.

The total number of subunits present in one receptor molecule is also unknown at present, but recent evidence in our laboratory suggests that it is five, as is the case with the nicotinic acetylcholine receptor. If, then (based upon the co-expression evidence), one takes all of the combinatorial possibilities for a pentamer containing $\alpha + \beta$, or $\alpha + \beta + \gamma$, or $\alpha + \beta + \delta$, or $\rho + \beta + \gamma$ types and if one allows that up to two different α isoforms or two β isoforms can co-exist, then the theoretical total of possible receptor subtypes is several thousand. Beyond that, alternative splicing of the RNA precursor is also known to occur in at least some isoforms (Whiting et al. 1990; Kofuji et al. 1991; Glencorse et al. 1991b). This would multiply by two for every such case the number of possible receptor subtypes. It is improbable that most of these theoretical possibilities actually occur in the brain, but the distribution patterns of the individual mRNAs are so complex that probably a number of combinations of at least between 50 and 100 would be compatible with it. The heterogeneity of the $GABA_A$ receptors in the brain is, therefore, at a level hitherto unheard of for any receptor. One must suppose that when a single signal (GABA) is used ubiquitously for fast inhibitory transmission in a vast number of neural circuits, differentiation between them is accomplished by great heterogeneity in its receptors, achieved by combinatorial alternatives.

This situation could be of benefit in pharmacotherapy. With sophisticated medicinal chemistry, it may be possible to distinguish many of these combinations as drug targets, and hence achieve selective actions on particular loci or cells that are affected in particular disorders.

Roles of $GABA_A$ Receptors in Brain Disorders

With so many receptor subtypes and such a general utilisation of GABA in the human nervous system, it is to be expected that this receptor type will be a target for a number of clinically important drugs. For example, the activity of anxiolytic compounds such as the benzodiazepines, and the anxiogenic β-carbolines, both of which bind to this receptor when the γ2 subunit is present, indicate a role for these receptors in some forms of anxiety disorders. The anticonvulsant activity of barbiturate drugs, which also bind to this receptor, similarly suggests the involvement of this complex in epileptic syndromes. An excess of neuronal inhibition may lead to clinical depression and there have been some indications for a possible involvement of this receptor in bipolar or unipolar depressive states.

Each of the isoforms noted above (apart from the alternative splice variants) has been shown to be from a different gene, so that at least 17 genes specify $GABA_A$ receptors. This, together with the ubiquity of this receptor class in the brain, renders it reasonable to regard these as candidate genes for one or other forms of the disorders noted above, or perhaps some others.

As an example, a strong genetic element (a single, autosomal dominant locus, with incomplete penetrance) has been described in panic disorder, in the pedigrees studied by Crowe et al. (1987) and Pauls et al. (1980), and twin studies have reported a concordance rate of about 40% in monozygotic twin pairs compared to 0%–4% in dizygotic twins (Torgersen 1983). Since benzodiazepines can provide relief in panic attacks in many cases and their antagonist, flumazenil, provokes these attacks in (and only in) the patients (Nutt et al. 1990), a genetic locus for one of the components of the GABA system, which may be a receptor subunit gene, is therefore well worth investigation.

Application of GABA Receptor Genes as Linkage Probes

When pedigrees are collected by clinicians in which a neurological or psychiatric disease appears to be vertically transmitted genetically, and if the GABA-ergic system seems, from mechanistic or therapeutic considerations, to be a plausible candidate for the causative locus, how would one test for its genetic segregation with the disease? Genes encoding three elements of the GABA-ergic system are now identified (Table 1): the $GABA_A$ receptor subunit family of genes, a gene for the enzyme specific to GABA synthesis in nerve terminals (GAD), and a gene for the GABA transporter for the re-uptake of GABA (Nelson et al. 1990). All three types should be used in such testing as this becomes feasible.

The first stage is to determine the human chromosomal localisation of each of the human genes, at the level of high-resolution chromosomal banding. A programme for this has been started in our laboratory with collaborators (Buckle et al. 1989). From this and similar studies made elsewhere, a number of these genes have been mapped (Table 1). A total of at least 19 gene loci would be available when this set is completed. Where any neuropsychiatric disorder from existing linkage evidence is known to map to the same band as one of these, the candidate gene approach is then well worth pursuing. Thus, the GABA receptor α3 subunit locus in Xq28 is in the region reported as a locus for an X-linked bipolar depressive illness, but the latter has been recently re-investigated and (with probes including one for that α3 locus) has been excluded in one set of pedigrees (van Broeckhoven et al. 1991). The β3 subunit locus is in a region on chromosome 15 which is deleted in a few patients with Angelman syndrome (Wagstaff et al. 1991), giving it tentative candidate status.

The second stage required is to screen DNAs from affected and nonaffected individuals, e.g. in family studies, with polymorphic probes for some of these 19 genes, to seek linkage or an association, analysed by one of the established genetic strategies. However, the $GABA_A$ receptor cDNAs have proved to be exceptionally stable in the population, presumably due to the vital functions involved, so that informative restriction fragment length polymorphisms (RFLPs) cannot be obtained. The same is true when genomic DNA is used in seeking the RFLP probes. For a case such as this, we must proceed to a third stage, the identification of "dinucleotide-repeat" polymorphisms by a special strategy based upon the polymerase chain reaction (Hicks et al. 1991). Using this method, we have obtained such polymorphisms, e.g. for the human α3 subunit locus (Hicks et al. 1991) with four alleles, informative in 36% of humans tested. Similar studies here are providing such probes for the α1

Table 1. Cloned genes for components of the GABA-ergic signalling system localised to human chromosomes

Gene	Chromosome	Reference
$GABA_A$ receptor α1	5 q34	Buckle et al. (1989)
α2	4 p13	Buckle et al. (1989)
α3	X q28	Buckle et al. (1989)
β1	4 p13	Buckle et al. (1989)
β3	15 q11-13	Wagstaff et al. (1991)
γ2	5	Unpublished results
δ	1 p	Sommer et al. (1990)
Glutamate decarboxylase	2	Erlander et al. (1991)

and some other subunit genes. The use of such (CA)-repeat polymorphisms renders genes which are very stable in the human population, such as those involved here, available for candidate gene screening. It is presently the method of choice for testing hypotheses linking mutations in a system such as that of GABA-ergic signalling to inherited neurological and psychiatric diseases, a scrutiny which is now proceeding in several collaborative studies.

References

Barnard EA, Darlison MG, Seeburg P (1987) Molecular biology of the $GABA_A$ receptor: the receptor/channel superfamily. Trends Neurosci 10:502-509

Buckle VJ, Fujita N, Ryder-Cook AS, Derry JMJ, Barnard PJ, Lebo RV, Schofield PR, Seeburg PH, Bateson AN, Darlison MG, Barnard EA (1989) Chromosomal localization of $GABA_A$ receptor subunit genes: relationship to human genetic disease. Neuron 3:647-654

Crowe RR, Noyes R Jr, Wilson AF, Elston RC, Ward LJ (1987) A linkage study of panic disorder. Arch Gen Psychiatry 44:933

Cutting GR, Lu L, O'Hara BF, Kasch LM, Montrose-Rafizadeh C, Donovan D, Shimada S, Antonarakis SE, Guggino WB, Uhl GR, Kazazian Jr H (1991) Cloning of the γ-aminobutyric acid (GABA) ρ1 cDNA: a GABA receptor subunit highly expressed in the retina. Proc Natl Acad Sci USA 88:2673-2677

Erlander MG, Tillakaratne NJK, Feldblum S, Patel N, Tobin AJ (1991) Two genes encode distinct glutamate decarboxylases. Neuron 7:91-100

Glencorse TA, Bateson AN, Darlison MG (1991) Sequence of the chicken $GABA_A$ receptor γ2-subunit cDNA. Nucleic Acids Res 18:7157

Glencorse TA, Bateson AN, Darlison MG (1992) Differential localization of two alternatively spliced γ2 mRNAs in the chick brain. Eur J Neurosci (in press)

Hicks AA, Johnson KJ, Barnard EA, Darlison MG (1991) Dinucleotide repeat polymorphism in the human X-linked $GABA_A$ receptor α3-subunit gene. Nucleic Acids Res 19:4016

Kofuji P, Wang JB, Moss SJ, Huganir RL, Burt DR (1991) Generation of two forms of the γ-aminobutyric $acid_A$ receptor γ2-subunit in mice by alternative splicing. J Neurochem 56:713-715

Luddens H, Wisden W (1991) Function and pharmacology of multiple $GABA_A$ receptor subunits. Trends Pharmacol Sci 12:49-51

Nelson H, Mandiyan S, Nelson N (1990) Cloning of the human brain GABA transporter. FEBS Lett 269: 181-184

Nutt DJ, Glue P, Lawson C, Wilson S (1990) Flumazenil provocation of panic attacks. Arch Gen Psychiatry 47:917-925

Olsen RW, Tobin AJ (1990) Molecular biology of $GABA_A$ receptors. FASEB T 4:1469-1480

Pauls DL, Bucher KD, Crowe RR, Noyes R (1980) A genetic study of panic disorder pedigrees. Am J Hum Genet 32:639-644

Sigel E, Baur R, Trube G, Möhler P, Malherbe P (1990) The effect of subunit composition of rat brain $GABA_A$ receptors on channel function. Neuron 5:703-711

Sommer B, Poustka A, Spurr NK, Seeburg PH (1990) The murine $GABA_A$ receptor δ-subunit gene: structure and assignment to human chromosome 1. DNA Cell Biol 9:561-568

Torgersen S (1983) Genetic factors in anxiety disorders. Arch Gen Psychiatry 40:1085-1089

Van Broeckhoven C, De Bruyn A, Raeymaekers P, Sandkuijl L, Hicks AA, Barnard EA, Darlison MG, Mendelbaum K, Mendlewicz J (1991) Exclusion of manic depressive illness from the chromosomal regions xq27-q28 and 11p15. HGM11, August, London, p 128

Wagstaff J, Knoll JHM, Fleming J, Kirkness EF, Martin-Gallardo A, Greenberg F, Graham JM, Menninger J, Ward D, Venter JC, Lalande M (1991) Localization of the gene encoding the $GABA_A$ receptor β3 subunit of the Angelman/Prader-Willi region of human chromosome 15. Am J Hum Genet 49:330-337

Whiting P, McKernon RM, Iversen LL (1990) Another mechanism for creating diversity in γ-aminobutyrate type A receptors: RNA splicing directs expression of two forms of γ2 subunit, one of which contains a protein kinase C phosphorylation site. Proc Natl Acad Sci USA 87:9966-9970

Wisden W, Morris BJ, Darlison MG, Hunt SP, Barnard EA (1988) Distinct $GABA_A$ receptor α subunit mRNAs show differential patterns of expression in bovine brain. Neuron 1:937-947

Discussion of Presentation E. A. Barnard

Ackenheil

You gave a good example for approaching the problem of using such candidate genes. You have mentioned the panic disorder as an inherited disease; maybe it is inherited, but whether this is really related to the GABAergic system seems questionable. You mentioned the results with the possible agonist flumazenil. But I have some doubts on this, because it could happen that pretreatment with other benzodiazepines may later on provoke the effects of flumazenil. There are other hypotheses concerning the catecholamine systems. My opinion is, panic disorder is not a very well defined illness for linkage studies.

Barnard

I would make two comments on that. Firstly, I have mentioned two different types of experiments, one of them with a partial inverse agonist, which produces severe anxiety, that is, in normal individuals who have never had any benzodiazepines before. In other words, this is the observation when you give this drug to normal individuals, a situation that has nothing to do with the inheritance of panic disorders. It is simply to say that a drug which acts on the GABA receptor in the opposite way of benzodiazepines produces severe anxiety, thus suggesting that there is a natural mediator there, which is normally increasing the action of GABA. In the case of the panic patients, the observation of Nutt et al. (1990) was that the benzodiazepine agonist flumazenil when given to these patients provoked a panic attack. There is no reason to believe that prior treatment with benzodiazepine was predisposing to this, since none of the patients had ever had benzodiazepines on a regular basis and all had been free of drugs for 3 months and had no withdrawal symptoms.

Ackenheil

We have one result: If you give one dose of a benzodiazepine to a normal person and withdraw the benzodiazepine after 2 days and give them flumazenil, then you can provoke anxiety; I would not say panic attacks.

Barnard

In untreated normal subjects, it is the b-carboline drug FG7142, an inverse agonist, that you need to provoke the anxiety attack. But there is no suggestion that flumazenil, the antagonist, provokes panic attacks in normal individuals. That is the difference: it does so in panic patients, but never in normal individuals. The evidence for the inheritance of panic disorder in some forms I must leave to the papers of Raymond Crowe (see the references I quoted) who has patient pedigrees which clearly show this. Further, recent studies of Steven Paul and colleagues at NIMH, USA, support this, since they showed that in these pedigrees a biological marker, high saccadic eye velocity after benzodiazepine treatment, is found to associate with the panic disorder and not with normals. If a biological marker follows the diagnosis in a pedigree, that does seem to provide further evidence for inheritance of the psychopathology.

Kidd

I do not really disagree with anything you have said, but given the effect, one should still be cautious about doing association studies with CA repeats, because they do have high mutation rates and that will tend to destroy any association that might exist with more stable mutational events. But, as you said that there are no polymorphisms in the gene, that method is basically all you have available.

Barnard

Yes, we are aware of that but firstly, the mutation probability in three generations seems to be very small, since we have seen no evidence for any without VA-repeat-polymorphisms. Nevertheless, to guard against it we try to use such polymorphisms at several different loci on the gene. It is quite easy to get these dinucleotide repeat polymorphisms at several different locis on the gene. Remember, the GABA receptor genes, which probably relate to that significance, are very large. All of the genes we have looked at are 100–200 kb or more in size, so they have got lots of opportunities for dinucleotide repeats and if you use two, you are not going to have a mutation in the same patients at two different loci.

Maier

Perhaps a more appropriate model for testing this GABA hypothesis would be hereditary epilepsy instead of panic disorders. Now, for the epilepsy there is an animal model in mice and recently there was a report in *Nature* or *Science* that for this animal model a locus has been identified on a chromosomal region which corresponds to chromosome 3 in humans. Unfortunately in your list of GABA subunits on the chromosomes there was no chromosome 3. Is it still hoped that you will identify such a GABA receptor subunit on chromosome 3?

Barnard

There are at least 17 GABA receptor genes and so far only about 8 of them have been mapped (see Table 1 in Barnard, this volume). Since they seem to be spread over many chromsomes, it is still quite possible for one to be on chromosome 3. Secondly, we have not looked at animal models of epilepsy, because there is some dispute as to whether they are really faithful correlates of human epilepsies. The studies on epilepsies being done in collaboration with groups in London are focusing on human pedigrees for some forms in which an inherited epilepsy seems pretty certain. Since there was no evidence for X-linkage in the forms studied, we are supplying them with the probes as we get them for the autosomal GABA receptor genes.

Gurling

I would like to make a point about CA block polymorphisms and evolutionary stability and hence the question of whether they are useful in association studies. We found 2 CA block polymorphisms in one clone, 7 kb apart, and showing strong linkage disequilibrium. Furthermore, the size of one repeat correlated with the size of the other repeat, 7 kb away. We assume that there was some mechanism that was creating length polymorphisms that coordinated the two linked variations. And furthermore, the strong evolutionary conservation of these repeats is useful for association studies. We subsequently heard that other research groups found linkage disequilibrium and an association between RFLPs and CA blocks and this should be useful for association studies.

Vogel

The only human type of epilepsy that has been mapped is the benign epilepsy with an autosomal dominant type of inheritance. It has been mapped to 20q13. Have you already tried 20q13 with GABA receptor gene probes?

Barnard

The benign familial neonatal convulsions form of epilepsy has been reported to be on chromosome 20, and the Unverricht-Lundborg myoclonic type on 21. We have no GABA receptor gene probes localizing to those chromosomes as yet, but it remains possible that one of the genes does.

Mallet

We have further evidence for the chromosome 20 location of that epilepsy.

Barnard

There are also some pedigrees which have been reported for juvenile myoclonic epilepsy. Therefore, there are several inherited epilepsies. We will have to test, when possible, whether one of the GABA receptor probes maps to the gene in question.

Gene Assignment by Random Search in Human Genome

L. Peltonen

Currently close to 2000 genes have been mapped to the human genome and over 500 genetic disorders have their chromosomal assignment (Human Gene Mapping Conference 1991). The speed of further assignments in years to come is predicted to increase since the tools for efficient mapping of any human disease are better than ever. Most disorders which have already been assigned to a chromosome were mapped by locating the wild-type gene. This was for the case phenylalanine hydroxylase, deficient in phenylketonuria, which was mapped to 12q actually independently of the disease. However, in the case of about 50 human mendelian diseases the clinical phenotype was mapped by family linkage studies using polymorphic anchor markers localized in individual chromosomes. Previously these markers represented protein polymorphisms, but more recently the accumulated information on human genome has revealed an enormous wealth of DNA polymorphisms and, consequently, a much more informative map of markers has emerged (Botstein et al. 1980). Table 1 provides a list of examples of mapped mendelian disorders for which the biochemical basis was not previously known and for which mapping really served as the first step towards a basic understanding of the disease (McKusick 1991).

When studying a clinical phenotype by linkage analyses, several criteria can be set for a disease which is ideal for mapping: A single affected locus should be assumed in the analyzed families, the diagnosis should be well defined, and the number of wrong phenocopies should be minimal. Further, affected individuals should preferentially be found in more than one generation. These strict criteria would exclude most fatal and recessive diseases from linkage analyses. However, linkages have also been reported for such diseases but the fact remains that there is a relatively low number of mapped recessively inherited clinical phenotypes when compared to dominantly inherited conditions. The explanation for success in most cases has been a large number of families, a unique population structure or, most importantly, the very recent development of highly informative polymorphisms which extract maximal information from small pedigrees.

The polymorphisms, small differences in DNA between individuals, are spread throughout the human genome. Sometimes they appear in the form of the presence or absence of restriction endonuclease recognition sites (restriction fragment length polymorphisms, RFLPs) or they can represent a variation in the number of tandem repeat nucleotide sequences at a particular region of a chromosome. Two kinds of such repeat length polymorphisms have been observed, their main difference being the length of the repeat unit (Nakamura et al. 1987; Tautz et al. 1986). Variable

Table 1. Diseases mapped by linkage analyses

Disease	Assignment
Huntington disease	4p
Facioscapulohumeral muscular dystrophy	4q
Spinal muscular atrophy, several types	5q
Adenomatous polyposis of colon	5q
Juvenile myoclonic epilepsy	6p
Progressive myoclonic epilepsy	21q
Spinocerebellar ataxia (one form)	6p
Cystic fibrosis	7q
Friedreich ataxia	9q
Torsion dystonia	9q
Tuberous sclerosis	9q, 11q
Ataxia-telangiectasia	11q
Marfan syndrome	15q
Polycystic kidney disease	16p
Batten disease	16p
Neuronal ceroid lipofuscinosis, infantile form	1p
Familial osteoarthrosis	12q
Neurofibromatosis	17q
Myotonic dystrophy	19q
Alzheimer disease (one form)	21q
Acoustic neuroma, bilateral	22q

number of tandem repeats (VNTRs) have units from 11 to 60 base pairs, whereas so called short tandem repeats (STRs) are most frequently di-, tri- or tetranucleotide repeats. The repeat polymorphisms represent a "new generation" of markers since they reveal 5 - 20 or even more alleles in the population whereas a typical RFLP polymorphism reveals only two. When the new multiallelic markers instead of two-allelic markers are used in linkage analyses and one tries to trace the segregation of polymorphic loci through generations, the effect on linkage analysis is comparable to a change from a black and white to a multicolor picture. In each mating it is possible to tell exactly how alleles were inherited, and this makes even small families highly informative.

Most of the RFLPs and VNTRs are detected using hybridization with the labeled probe which visualizes the alleles from the restriction digested DNA, electrophoresed, and transferred to the membrane. The method is slow and complicated and requires a significant amount of manual labor, thus being expensive. It is not surprising that considerable effort in human gene mapping is being invested to develop these polymorphisms to an amplifiable form (Saiki et al. 1985). Polymerase chain reaction (PCR) amplifies a given DNA segment and magnifies the signal of this particular locus so much that hybridization becomes unnecessary but the alleles can be directly visualized with ethidium bromide or silver staining after the alleles are electrophoretically separated in agarose or, preferentially, in polyacrylamide gels. Detection of STR polymorphisms is always based on PCR, the disadvantage of these markers being that due to a very small size difference between the alleles, electrophoresis has to be carried out in denaturing (urea containing) sequencing gels to confirm correct identification of alleles.

Since the Human Gene Mapping XI Congress in London it is obvious that the marker map of the future, immediately applicable to disease mapping, will be based on multiallelic, amplifiable polymorphisms. Unfortunately these polymorphisms seem to be unevenly situated in individual chromosomes. An ideal marker set of, for example, 600 markers at 5-cM intervals covering the genome cannot be obtained only based on these highly informative polymorphisms. The gaps in this map have to be filled with biallelic, classical RFLPs. However, these loci, too, can be amplified based on PCR and instead electrophoretic separation, novel techniques such as solid-phase minisequencing (Syvänen et al. 1990) or oligonucleotide ligase assay (Nickerson et al. 1990) can be used to identify the alleles. All steps in these systems can be automatized and a rapid detection of 20-50 polymorphisms in a significant amount of samples can be carried out fast and efficiently.

To conclude, a highly developed, technically user-friendly marker map is almost reality and available for efficient linkage analyses of any inherited disease for which representative families can be found. The application of this more complete genomic map together with new techniques facilitating rapid detection of alleles and improved linkage analyses should also be highly useful in the mapping of more complex human phenotypes and polygenic traits.

References

Botstein D, White RL, Skolnick M, Davis RW (1980) Construction of a genetic linkage map in man using restriction length polymorphisms. Am J Hum Genet 32:314-331

Human Gene Mapping XI Conference, London (1991) Cytogenet Cell Genet (in press)

McKusick VA (1991) Current trends in mapping human genes. FASEB J 5:12-20

Nakamura Y, Leppert M, O'Connell P, Wolff R, Holm T, Culver M, Martin C, Fujimoto E, Hoff M, Kumlin E, White R (1987) Variable number of tandem repeat (VNTR) markers for human gene mapping. Science 235:1616-1622

Nickerson DA, Kaiser R, Lappin S, Stewart J, Hood L, Landegren U (1990) Automated DNA diagnostics using an ELISA¬based oligonucleotide ligation assay. Proc Natl Acad Sci USA 87:8923-8927

Saiki RK, Scharf S, Faloona F, Mullis KB, Horn GT, Erlich HA, Arnheim N (1985) Enzymatic amplification of beta-globin genomic sequences and restriction site analysis for diagnosis of sickle cell anemia. Science 230:1350-1354

Syvänen A-C, Aalto-Setälä K, Harju L, Kontula K, Söderlund H (1990) A primer-guided nucleotide incorporation assay in the genotyping of apolipoprotein E. Genomics 8:684-692

Tautz D, Trick M, Dover GA (1986) Cryptic simplicity in DNA is a major source of genetic variation. Nature 322:652-656

Discussion of Presentation L. Peltonen

Gurling
I would like to ask you about using your solid-phase minisequencing method in comparison to the dinucleotide-trinucleotide microsatellite system. Do you think it can be applied in analyzing these microsatellites as well?

Peltonen
No, I think it is too complicated for this system, which is ideal for just one base pair. We have analyzed deletions and insertions, but currently dinucleotide or trinucleotide repeat is too complicated for this assay. This holds for oligonucleotide ligase assay as well. These systems show their greatest strength in classical RLFPs and the results can be repeated beautifully from one assay to the other.

Family Studies of Affective Disorders with 11p15.5 DNA Markers

J. Mallet, A. Malafosse, M. Leboyer, T. d'Amato, S. Amadéo, M. Abbar, M.-C. Babron, D. Campion, O. Canseil, D. Castelnau, A. DesLauriers, F. Gheysen, B. Granger, B. Henriksson, H. Loo, M.-F. Poirier, O. Sabaté, D. Samolyk, E. Zarifian, and F. Clerget-Darpoux

Family, twin, and adoption studies suggest the involvement of genetic factors in manic-depressive illness (MDI) but the mode of inheritance remains elusive. The remarkable success with which the classical logarithm of odds (lod) score method has contributed to the localization of loci for numerous diseases has led many authors to apply this strategy to complex illnesses such as affective disorders. Most pertinent was the study conducted by Egeland et al. in 1987, which was based on an extended pedigree in order to minimize a possible genetic heterogeneity. These investigators reported a linkage between MDI and insulin (INS) and c-Harvey-ras-1 oncogene (HRAS) markers located within 11p15. But a recent reinvestigation of the Amish pedigree and the study of a set of non-Amish pedigrees call into question this initial finding (Kelsoe et al. 1989).

There is, however, no compelling argument to reject the 11p15.5 locus in considering the pitfall attendant upon the use of this parametric method in complex diseases. Linkage data only suggest that there is no single major gene at this locus conferring susceptibility to MDI. These considerations together with the presence within this locus of the gene encoding tyrosine hydroxylase (TH), the rate-limiting enzyme for catecholamine synthesis, led us to reinvestigate this region with nonparametric methods in the population and family samples.

Population Studies

Using a sample of 50 patients and 50 controls we initially reported an association between Taq I and Bgl II alleles in the 5' and 3' region of the TH gene, respectively (Leboyer et al. 1990). We have now extended this study by examining a larger sample (100 probands), and the data provide further weight to our initial finding. Two other studies have shown a similar trend in the frequency of the 3' allele without having reached, as yet, statistical significance (Körner et al. 1990; Gil et al. 1992). Nevertherless, pooling of these data with respect to the 3' allele strengthens our results ($p < 0.001$).

Family Studies

Families were first analyzed by the affected pedigree member (APM) method recently developed by Weeks and Lange (1988) to detect possible deviations from independent segregation between marker loci and MDI. The APM method offers a decisive advantage over parametric methods in requiring neither specification of the mode of transmission nor the use of estimated parameters. It was selected instead of the classical sib-pair method which would have required a sample larger than that currently available to us.

Collective analysis of our 11 French families (137 individuals including 81 affected members) with the above method shows that the segregation of the disease TH, INS and HRAS markers as well as the corresponding haplotypes is not independent. The highest statistical significance was obtained for the haplotypes with a p value less than 10^{-6} for TH-INS-HRAS1 haplotype.

Analysis of the families individually indicates, however, that not all of them contribute to the deviation from independent segregation, some being clearly negative. Such differences are most likely to reflect a genetic heterogeneity. Most interestingly, evidence of deviation from nonrandom segregation was also found when the typing data from previously published pedigrees were reanalyzed by the APM method. As with the French pedigrees, there is evidence for genetic heterogeneity. This phenomenon was even shown within the large Amish pedigree where non-independent segregation was statistically significant with the left extension of the pedigree.

The observation that the same alleles at the three studied loci appear to segregate nonindependently with affective disorders in some families suggests the existence of linkage disequilibrium between these alleles and the disease. This supposition was confirmed by applying the method of Boehnke to our family data.

Taken together, the above findings support the conclusion that there exists a DNA sequence within the TH-INS-HRAS1 region, which in concert with other gene(s) and/or environmental factors causes a predisposition to affective disorders.

Regulatory Mechanisms of TH Gene Expression

While there is no evidence that the sequence which confers vulnerability to MDI is associated with the TH, the gene does stand as a good candidate in view of the catecholamine hypothesis of affective disorders. Moreover, an elaborate array of biochemical mechanisms have been shown to modulate its activity in response to environmental factors as anticipated for a gene possibly involved in the physiopathology of affective disorders. Changes in TH activity may result from short-term as well as long-term mechanisms. The former arise from phosphorylation events, and several residues have been shown to be phosphorylated in vitro and in vivo by multiple kinases, including the cyclic AMP-dependent protein kinase (PKA) and the calmodulin-dependent protein kinase II (CaM-PKII; Cambell et al. 1986; Haycock 1990).

Long-term changes have mostly been studied in the context of transsynaptic stimulation of TH activity. This phenomenon, which is elicited by a number of conditions, including electrical stimulation, cold stress, or treatment with reserpine, causes an increase in the activity of the enzyme in the adrenal medulla, the superior cervical ganglia, and noradrenergic cells of the locus coeruleus. The change in TH enzyme activity may last several weeks and the underlying mechanisms could form the basis of an information storage system (Black et al. 1984, 1987; Comb et al. 1987). Transsynaptic stimulation of TH activity is due to an increase in the number of enzyme molecules as a consequence of an increase in TH mRNA (Black et al. 1985; Faucon Biguet et al. 1986), which is likely to involve both transcriptional and posttranscriptional events. Recently we have established that a sequence located 200/207 bp from the transcriptional start site plays a critical role in the establishment of this long-term effect, thereby providing a clue to study the mechanisms that translate nerve activity into gene activity.

In addition, in humans, the TH gene generates through alternative splicing events four different mRNAs, designated HTH mRNA-1, -2, -3, and -4 (Grima et al. 1987; Le Bourdellès et al. 1988; Kaneda et al. 1987). HTH-2 mRNA and HTH-3 mRNA differ from HTH-1 mRNA by an insertion at position 90 of 12 or 81 nucleotides, respectively, while HTH-4 mRNA contains both insertions. At the protein level, the corresponding diversity is restricted to the N-terminal, the regulatory domain of the enzymes. By taking advantage of the *Xenopus* oocyte expression system, we initially established that these isoforms display significant differences in their specific activities, HTH-1 and HTH-2 being the most and least active forms, respectively (Horellou et al. 1988). These findings were later confirmed by transfection experiments in COS cells (Kobayashi et al. 1988). Recently, we have produced HTH-1 and HTH-2, individually, in bacteria. Both isoforms could then easily be purified to homogeneity, thereby allowing a detailed analysis of the sites that are being phosphorylated by various kinases and how these events, which subserve short-term response to presynaptic stimuli, affect their enzymatic characteristics. In particular, we showed that the presence of four additional amino acids in HTH-2 provides a consensus sequence which allows the adjacent serine to be phosphorylated by CaM-PKII. Thus, HTH-2 possesses an additional site of phosphorylation relative to HTH-1, a characteristic which provides a certain degree of diversity in the mechanism underlying the short-term regulation of the enzymes (Le Bourdellès et al. 1991).

All four HTH mRNAs are expressed in pheochromocytoma tumors. In adrenal glands, HTH-4 appears to be lacking while in the brain HTH-1 mRNA and HTH-2 mRNA predominate. The presence in the same tissue of several types of HTH mRNA which encode enzymes with distinct characteristics raises the possibility that alternative splicing plays a role in the regulation of catecholamine biosynthesis in normal and pathological neurons in addition to transcriptional and posttranslational events.

Clearly, mutations within the TH gene which affect the regulation of its expression may alter the response of the gene to nerve activity. It should be noted that some regulatory sequences may reside a long distance from the coding and may in fact be closer to the coding sequence of a neighbor gene than to that of the TH itself.

Work in progress to pinpoint the sequence in the TH, INS, HRAS region that confer vulnerability to MDI will reveal whether it affects the expression of the TH gene itself.

References

Black IB, Adler JE, Dreyfus CF, Jonakait GM, Katz DM, LaGamma EF, Markey KM (1984) Neurotransmitter plasticity at the molecular level. Science 225:1266-1270

Black IB, Chikaraishi DM, Lewis EJ (1985) Trans-synaptic increase in RNA coding for tyrosine hydroxylase in rat sympathic ganglion. Brain Res 339:151-153

Black IB, Adler JE, Dreyfus CF, Freidman WF, La Gamma EF, Roach AH (1987) Biochemistry of information storage in the nervous system. Science 236:1263-1268

Cambell DG, Hardie DG, Vulliet PR (1986) Identification of four phosphorylation sites in the N-terminal region of tyrosine hydroxylase. J Biol Chem 261:10489-10492

Comb M, Hyman SE, Goodman HM (1987) Mechanisms of transsynaptic regulation of gene expression. TINS 10:473-478

Egeland J, Gerhard D, Pauls D, Sussex J, Kidd K, Allen C, Hostetter A, Housman D (1987) Bipolar affective disorders linked to DNA markers onchromosome 11. Nature 325:783-787

Faucon Biguet N, Buda M, Lamouroux A, Mallet J (1986) Time course of the changes of TH mRNA in rat brain and adrenal medulla after a single injection of reserpine. EMBO J 5:287-291

Gill M, Castle D, Hunt N, Clements A, Sham P, Murray RM (1992) Tyrosine hydroxylase polymorphisms and bipolar affective disorder. J Psychiatr Res (in press)

Grima B, Lamouroux A, Boni C, Julien JF, Javoy-Agid F, Mallet J (1987) A single human gene encodes multiple tyrosine hydroxylases predicted to differ in their functional characteristics. Nature 326:707-711

Haycock JW (1990) Phosphorylation of tyrosine hydroxylase in situ at serine 8, 19, 31, and 40. J Biol Chem 265:1-10

Horellou P, Le Bourdellès B, Clot-Humbert J, Guibert B, Leviel V, Mallet J (1988) Multiple tyrosine hydroxylase enzymes, generated through alternative splicing, have different specific activities in *Xenopus* oocytes. J Neurochem 51:652-655

Kaneda N, Kobayashi K, Ichinose H, Kishi F, Nakazawa A, Kurosawa Y, Fujita K, Nagatsu T (1987) Isolation of a novel cDNA clone for human tyrosine hydroxylase: alternative mRNA splicing produces four kinds of mRNA from a single gene. Biochem Biophys Res Commun 146:971-975

Kelsoe J, Ginns E, Egeland J, Gerhardt D, Goldstein A, Bale S, Pauls D, Long R, Kidd K, Conte G, Housman D, Paul S (1989) Re-evaluation of the linkage relationship between chromosome 11p loci and the gene for bipolar affective disorder in the Old Order Amish. Nature 342:238-243

Kobayashi K, Kiuchi K, Ishii A, Kaneda N, Kurosawa Y, Fujita K, Nagatsu T (1988) Expression of four types of human tyrosine hydroxylase in COS cells. FEBS Lett 238:431-434

Körner J, Fritze J, Propping P (1990) RFLP alleles at the tyrosine hydroxylase locus: no association found to affective disorders. Psychiatry Res 32:275-280

Le Bourdellès B, Boularand S, Boni C, Horellou P, Dumas S, Grima B, Mallet J (1988). Analysis of the 5' region of the human tyrosine hydroxylase gene: combinatorial patterns of exon splicing generate multiple regulated tyrosine hydroxylase isoforms. J Neurochem 50:988-991

Le Bourdellès B, Horellou P, LeCaer J, Denèfle P, Latta M, Haavik J, Guibert B, Mayaux JF, Mallet J (1991) Phosphorylation of human recombinant tyrosine hydroxylase isoforms 1 and 2: an additional phosphorylated residue in isoform 2, generated through alternative splicing. J Bio Chem 266:17124-17130

Leboyer M, Malafosse A, Boularand S, Campion D, Gheysen F, Samolyk D, Henrikkson B, Denise E, Des Lauriers A, Lépine JP, Zarifian E, Clerget-Darpoux F, Mallet J (1990) Tyrosine hydroxylase polymorphisms associated with manic-depressive illness. Lancet 335:1219

Weeks DE, Lange K (1988) The affected pedigree member method of linkage analysis. Am J Hum Genet 42:315-326

Discussion of Presentation J. Mallet et al.

Mikoshiba

I would like to ask about the phosphorylation site. Also in the case of ß-receptors, there are two phosphorylation sites; the N-terminal site in the central nervous system is more highly phosphorylated and the C-terminal region is highly phosphorylated in peripheral tissues. According to your data there are obviously four phosphorylation sites. Is there any specific kinase that phosphorylates each site?

Mallet

Clearly, kinase A phosphorylates serine 40 in both isoforms while the calmodulin-dependent protein kinase II phosphorylates serines 19 and 40 in isoform 1 and serine 19, 31, and 40 in isoform 2 (Le Bourdellès et al. J Biol Chem 266:17124-17130, 1991). So definitely, some kinases phosphorylate specific sites. The question that we are now addressing concerns the tissue specificity of these phosphorylation events.

Mikoshiba

Is there any functional difference?

Mallet

Yes. For instance, the phosphorylation by the calmodulin-dependent protein kinase II reduces dopamine inhibition of isoform 2 but not of isoform 1 (Le Bourdellès et al. J Biol Chem 266:17124-17130, 1991).

Gurling

The question I want to ask concerns the appropriate form of analysis for looking for association when you have family data. Spurred on by Jacques Mallet's interesting finding with tyrosine hydroxylase, we went back to our original insulin, HRAS, and tyrosine hydroxylase data in Icelandic families with manic- depressive illness. We subsequently have analyzed more large kindreds which are negative on chromosome 11. We reanalyzed the data using an extended sib-pair approach and found them to be negative. Which is the best form of analysis for these type of data? Another question is, if you use the Weeks and Lange method you have two possibilities of testing for the involvement of the gene or that locus: one is the identity by the state and the other is identity by descent, which includes the whole pedigree. Identity by state includes just the information from the sibs. How should psychiatrists best proceed with this problem?

Mallet

To your first question. When we reinvestigate all the pedigrees, there is clearly a heterogeneity. The positive results come only from a set of the pedigrees. When dealing with a large pedigree it is important to dissect it into subpedigrees to see which gene is involved in which part of the pedigree, because of the heterogeneity within a pedigree. And this is the same with the Amish. The identity by descendent which correponds to the sib-pair analysis would have required a larger sample than with the Weeks and Lange method, which considers the identity by state.

Mendlewicz

You said that when you analyze other pedigrees which have a negative linkage score, the global results using the Week and Lange method in fact reveal positive results.

Mallet

We have analyzed all the pedigrees about which we have published reports. I did not say that each was positive, but that combining the data yields positive statistics. There is obvious heterogeneity.

Maier

We heard in the presentation of Dr. Van Broeckhoven how a lod score emerging from an affected-only pedigree analysis can drop when using the classical lod score method. If you applied the classical lod score to your data where is your lod score?

Mallet

We did apply the lod score method but the results were negative. I think the lod score method is fantastic, but it might not be the right approach when dealing with complex diseases. It might work in some extreme cases if you find a pedigree with a single major locus, but just because the lod score drops from 4 to 2 when extending a pedigree does not mean that there is nothing in that pedigree.

Ott

I would like to answer Hugh Gurling's question. I think there are several ways one can test. All these tests are valid, the Weeks and Lange method and the affected sib-pair method. They all test the same hypothesis, absence of linkage, but they do it in different ways. One should know which one is the most powerful in a particular situation. Then it depends on the correct model. I think all of them are correct. I have a question regarding this heterogeneity. I am not exactly sure how you did the analysis. But if, for example, you have no linkage between a trait and the marker and you look at those families, many of them will show positive lod scores. So, if one picks those with positive lod scores, presumably they will add up to a significant lod score. I think before one does this, one should be able to show that there is significant heterogeneity.

Mallet
We did the analysis with Francoise Clerget-Darpoux. The data have been pooled because the test allowed it.

Reich
I think it is important to clarify the situation. I would like to make some comments on the lod score method, implying that there is only a single locus. The lod score method is probably the best and most powerful method, if you have the correct model. There are two locus models now available for studying the inheritance of a trait where the trait can be marked by one locus. It seemed to me that might be the most appropriate way since your hypothesis is that there is at least one other locus, perhaps a major locus. Then in fact the tyrosine hydroxylase allele may be the penetrance locus. So, that program does exist, but have you compared it with the Weeks and Lange method?

Mallet
I agree with you and I know that there are lod score methods for two loci. But they remain to be tested.

McGuffin
Just a comment which I hope will provide clarification. I think that pooling the samples is probably the worst method of calculating when you have differing population rates of various alleles. And have you also been able to characterize hetereogeneity?

Mallet
We of course performed a heterogeneity test. Pooling the data yields positive results, but looking at each family individually, some contribute and some do not. So clearly there is heterogeneity.

Mendlewicz
If you look at the allelic distribution in the different cohorts, what are your results?

Mallet
The allelic distributions in the different pedigrees were very similar and the same alleles were affected.

Cloninger
I just wanted to comment that finding an association without linkage has a parallel in other disorders. Recently we have been able to reject linkage for alcoholism and DRD2 locus, but in looking at the six published or in press case control studies, the level of significance is now at the level of $p = 10^{-7}$, with alcoholics carrying the A1 allele in 45% of cases and nonalcoholics carrying it in 26% of cases. So I think we may encounter this parallel result not infrequently with complex disorders.

Kidd
One comment about Bob Cloninger's comment on the significance level. I would say that some of the studies Bob Cloninger mentioned should not have been pooled because the control samples are a bit heterogeneous.

Cloninger
We did a heterogeneity test which allowed for the B allele but not for the C1. We have been very careful in that.

Kidd
The point I wanted to make about your study was that the haplotype gave a much stronger result. That is a very important phenomenon, because haplotypes should give much stronger associations. In fact, as I mentioned in my talk, in PKU the individual restriction sites show no association, but the haplotypes do and it is very much the same in cystic fibrosis. So I think that is a very strong point. The only concern I have is that part of the haplotype involved HRAS, which is so far away that I would not expect that there is still disequilibrium. But there might be, depending on the nature of the underlying association. I would like to see more of the RFLPs in that region.

Propping
I just wanted to comment that the association did not relate to the haplotypes, but only to the restriction site. Further, you pooled the association data from your own sample, from our data, and from Michel Gill's data. And that makes three samples. In our sample the association was completely negative.

Mallet
It was statistically negative but followed the same trend.

Propping
In Michel Gill's data it was the other way around.

Mallet
Michel Gill's data also follow the same trend. Now coming back to your question. Of course one cannot do haplotypes in classical association studies. The haplotype I showed concerned family studies.

Vogel
Coming back to the association studies, they were population association studies, not family studies. When there are three series and the outcomes are as different as Dr. Propping is suggesting, then the decisive point is the heterogeneity test. If you follow Dr. McGuffin's suggestions, using Wolfe's method, then this is a modification of the chi-square test specifically adapted to this problem. When you use Wolfe's method, then you can be sure to identify heterogeneity.

Mallet
We did apply a heterogeneity test, allowing us to pool the data at least for one of the alleles.

Clinical Phenotypes: Problems in Diagnosis

M. T. Tsuang, M. J. Lyons, and S. V. Faraone

Looking back at the proceedings of our last conference, I saw that I had discussed unreliability in psychiatric diagnoses. However, the primary problem in psychiatric diagnosis no longer can be said to be unreliability. Fortunately, substantial and impressive progress has been made during the past two decades. We now face a more complex and perhaps less tractable issue — the validity of diagnosis. One reason to be less sanguine about progress in improving validity compared to reliability is that, while we can reliably measure reliability, identifying a clear-cut criterion for validity is problematic. If one wants to improve something, it is helpful, if not necessary, to be able to measure it.

When validity is considered, the question of "valid for what" must be addressed. Robins and Guze (1970) described a number of possible validators for psychiatric diagnosis such as outcome and laboratory tests. Given the topic of this conference, it seems appropriate to emphasize a genetic criterion for diagnostic validity — the presence of the relevant genotype (while acknowledging that it is only one of a number of possible criteria). Until relatively recently, the genotype remained a hypothetical construct, but recent developments in molecular genetics have made a concrete description of the genotype accessible. In the future, we may be able to validate a clinical diagnosis by determining whether it is associated with the presence of the pathogenic genotype.

To parallel the discussion of sources of unreliability during the last meeting, I would like to discuss the threats to diagnostic validity, or, in a sense, sources of "invalidity" in our data. These factors could also be considered reasons for misclassification or problems of case definition. These threats to diagnostic validity are: (a) etiological heterogeneity / phenocopies; (b) incomplete penetrance of the genotype; (c) variable expressivity of the genotype; and (4) pleiotropy. Figure 1 illustrates how these phenomena relate to validity and the following section briefly presents each of these issues.

Threats to Diagnostic Validity

Etiological Heterogeneity / Phenocopies

In genetic studies one is likely to encounter some clinical cases that are caused by some nongenetic or alternative genetic mechanism. Such cases become false positives because they do not possess the criterion genotype. The inconsistent and

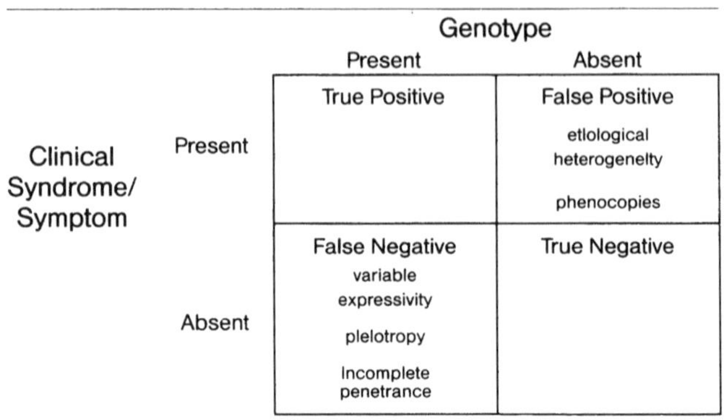

Fig. 1. Threats to diagnostic validity

sometimes conflicting results of mathematical modeling and linkage studies have most often been interpreted to indicate that psychiatric disorders are genetically and phenotypically heterogeneous. In other words, there may be several forms of the various disorders, each having a different mode of genetic transmission. Failure to recognize potential heterogeneity in the genetic mechanisms underlying psychiatric disorders creates fertile ground for conflicting findings. The hypothesis of genetic heterogeneity is appealing; it explains inconsistent results with a genetic hypothesis that has been substantiated in other areas of medical genetics. For example, Morton (1956) demonstrated that linkage between elliptocytosis and the Rhesus blood group could be conclusively demonstrated in some families and rejected in others. Ott (1985) discussed the case of Charcot-Marie-Tooth disease, a hereditary motor and sensory neuropathy. Although this disease appears clinically homogeneous, some families demonstrate linkage to the Duffy blood group while others do not. Glycogen storage diseases result in abnormal amounts of glycogen deposited in one or more organs. Ten different forms of the disease have been isolated, each corresponding to a different enzyme defect. Although the clinical manifestations of the ten subtypes are similar, each can be considered to be a different single major locus disorder (Vogel and Motulsky 1986).

An example of the problem of phenocopies is the status of individuals with unipolar depression in genetic studies of bipolar disorder. Dr. Deborah Blacker and I have done some work on this topic (Blacker and Tsuang 1991) and my comments are based on that work. A basic question is, when individuals with unipolar disorder are found in a bipolar pedigree, should they be classified as affected? Based on differences in sex ratios, age at onset, frequency of episodes, and severity of depressive episodes, there is substantial support for a distinction between unipolar and bipolar mood disorder (although there is some support for a single continuum of mood disorder; Tsuang and Faraone 1990). However, it is also highly likely that some meaningful subset of individuals with bipolar disorder experience only episodes of depression. Especially for individuals who have not lived through the entire risk period, the presence of only depressive episodes is not definitive evidence that the individual does not have bipolar disorder. Individuals with only depressive episodes have been counted as affected in a number of studies of bipolar disorder (e.g., Egeland et al. 1987; Detera-Wadleigh et al. 1987; Hodgkinson et al. 1987; Baron et al. 1987). There is a relatively high prevalence of major depression in the general population, and there are probably numerous etiological mechanisms producing depression. Therefore, it is very likely that some depression in bipolar pedigrees is unrelated to bipolar disorder.

The rate of bipolar disorder in bipolar pedigrees is relatively low (generally lower than the rate of unipolar depression), and a number of other clinical phenotypes have been included as affected by various investigators. For example, schizoaffective disorder (e.g., Rice et al. 1987a), unipolar depression (e.g., Cassano et al. 1989), cyclothymia (e.g., Akiskal et al. 1977), bipolar II (e.g., Endicott et al. 1985), alcoholism (e.g., Egeland and Hostetter 1983), and some personality disorders (e.g., Akiskal et al. 1977; Cassano et al. 1989) have been included as affected clinical phenotypes. Because of the great variability in depression, shading subtly from nonpathological dysphoric mood to profound and psychotic depression, prevalence estimates vary greatly depending upon the diagnostic criteria employed.

When a case of unipolar depression is observed in a bipolar pedigree, there are several possibilities: (1) the individual may have an illness that is entirely unrelated to bipolar disorder; (2) the individual may be expressing the bipolar genotype but will never manifest a manic episode; or (3) the individual may subsequently experience a manic episode and receive a bipolar diagnosis. Blacker and I have suggested a strategy for trying to identify unipolar conditions in individuals who are most likely to have the bipolar genotype by carefully studying "switchers," i.e., those who have had one or more depressive episodes before their first manic episode. In our Iowa sample, 12.5% of persons with unipolar conditions experienced a manic episode during a 30- to 40-year follow-up (Tsuang et al. 1981). Winokur and Wesner (1987) reported that depressive individuals who went on to have a manic episode were younger, more likely to be male, and had more severe depressive episodes compared to depressive individuals who never had a subsequent manic episode.

There are a number of reasons why some individuals with a "bipolar genotype" never experience a manic episode. Those who experience depression may commit suicide and therefore not live long enough to experience a manic episode. It may be that there is some genetic or environmental factor that is necessary for the occurrence of a manic episode in addition to the putative bipolar genotype — in its absence the individual may only experience depressive episodes. Another issue that may confound the assessment of the clinical phenotype is denial of manic episodes and acknowledgement of only depressive episodes. Goodwin and Jamison (1990) pointed out that it is not uncommon for individuals to deny their manic episodes, and Andreasen et al. (1986) reported that informant information is often more sensitive for manic episodes than is self-report. This phenomenon increases the risk of false negatives in studies of bipolar disorder. We believe that the most reasonable course is to consider depressed individuals in bipolar pedigrees to have an affected phenotype. However, in the statistical analyses, the probability that such an individual really represents a phenocopy should be used to differentially weight such cases.

Incomplete Penetrance of the Genotype

A classic mendelian model does not adequately describe the genetic transmission of psychiatric disorders. For example, if schizophrenia were caused by a fully penetrant dominant gene, we would expect that 50% of the offspring of one schizophrenic parent would become schizophrenic. The observed value is much lower. If schizophrenia were caused by a fully penetrant recessive gene, we would expect 100% of the offspring of two schizophrenic parents to be schizophrenic. The observed value is 36.6%. Thus, we must be able to accommodate the existence of individuals with the genotype but without the clinical phenotype (false negatives) in our statistical analyses. Fortunately, we are able to vary the penetrance parameter in our analyses of linkage data to take account of this phenomenon.

Variable Expressivity of the Genotype

It seems clear that the genotypes that cause or predispose to psychiatric disorders within the pedigrees that we study do not invariably lead to the same phenotype as that seen in the proband. Penetrance describes the probability that the genotype will be manifested in the phenotype, while expressivity refers to the degree to which the genotype will be reflected in the phenotype and indicates that the occurrence of the clinical phenotype is not an all-or-none phenomenon. There may be gradations of being "affected." Clinically, we observe a wide range of severities for many psychiatric disorders. The variable expressivity of genotypes for psychiatric disorders may produce a spectrum of both clinical and nonclinical phenomena. Sole reliance on narrowly defined traditional clinical phenotypes may limit progress. Holzman and Matthysse (1991), discussing schizophrenia research, suggested that modifying the diagnostic criteria might be considered reshuffling and redealing the "same deck of symptom cards." Their suggestion to include nonclinical phenotypes may be quite helpful in psychiatric genetics.

It is probably in the area of schizophrenia research where the greatest amount of work has been done to attempt to characterize subthreshold clinical phenotypes. The concept of a "schizophrenia spectrum" is congruent with the notion of a schizophrenia genotype with variable expressivity. Current perspectives regarding familial pathology and the schizophrenia spectrum have focused on the aggregation of DSM-III personality disorders (PD) among biological relatives of schizophrenic patients. Interest has primarily been focused on the familial prevalence of three PDs: schizotypal, schizoid, and paranoid. Although two recent studies failed to find a higher rate of schizotypal PD (SPD) among relatives of schizophrenic probands (Coryell and Zimmerman 1989; Squires-Wheeler et al. 1989), there is substantial evidence that biological relatives of schizophrenics demonstrate "subthreshold" pathology in the form of SPD.

Results for schizoid and paranoid PDs have been somewhat more controversial and contradictory. Baron et al. (1985) have noted an increase in paranoid PD in the relatives of schizophrenic probands (7.3%) versus control probands (2.7%). How ever, their results have been criticized (Rogers and Winokur 1988) for artificially inflating estimates of paranoid PD in relatives on the grounds that the sample of RDC-defined schizophrenic probands may have included delusional disorder probands. It has been suggested, based on family studies, that schizophrenia and delusional disorder are not genetically related (Kendler et al. 1981; Winokur 1985). Also, Winokur demonstrated increased rates of paranoid traits in relatives of delusional disorder patients, but not in the relatives of schizophrenics. Kendler et al. (1985) have shown that rates of DSM-III paranoid PD are not increased in the relatives of schizophrenics, but are increased in relatives of delusional disordered probands compared to relatives of controls. Currently, then, there does not appear to be strong evidence favoring paranoid PD as a member of the schizophrenia spectrum, although the question should not be considered to be definitively answered.

Surprisingly, schizoid PD has received relatively little interest and examination despite its apparent similarities to traits cited in the "familial" literature as relevant to family member pathology in schizophrenia (e.g., social isolation, affective impoverishment, etc.). In part, this may be attributable to a historical shift in which "schizoid" personality came to be interpreted in a more generalized and psychodynamic fashion (Siever and Gunderson 1983). Kety et al. (1968, 1975) and the Danish adoption studies are most frequently cited as providing evidence that potentially refutes a schizoid-schizophrenia genetic link. Given the utilization of a rather general DSM-II definition of schizoid PD, it is difficult to know precisely how such individuals were diagnosed.

The findings of a replication in an adoption sample (Kety 1988) indicate findings similar to Kety and coworkers' earlier report. However, the findings of Baron et al. (1985), utilizing DSM-III criteria for schizoid PD, failed to detect significant differences, although the relatives of schizophrenics did show a higher rate of schizoid PD than control relatives (1.6% versus 0%). Of interest, however, is the fact that Baron and coworkers (1985) incorporated a "probable schizotypal" PD category that required only two of the necessary four DSM-III criteria. Compared to relatives of controls, relatives of schizophrenics demonstrated a significantly higher rate of probable SPD (12.1% versus 6.5%). Since a symptom overlap exists between the DSM-III criteria for schizotypal and schizoid PDs (e.g., affective constriction, social isolation), it is possible that family members who would otherwise have met criteria for schizoid PD were instead diagnosed as "probable schizotypal" PD. Finally, Kendler et al. (1984, 1985) have demonstrated an increased prevalence of a combined "schizotypal-schizoid" PD in biological relatives of schizophrenics, but their results do not allow for a distinction between "schizotypal" and "schizoid" traits. Thus, as with paranoid PD, strong evidence has yet to be presented for establishing a link with schizophrenia. Clearly, among Axis-II disorders, schizotypal PD is the strongest candidate for a relatively mild illness genetically related to schizophrenia.

Previous research indicates that negative symptoms are characteristic of the biological relatives of schizophrenic patients (Kendler et al. 1983). This suggests that the negative symptoms in relatives of schizophrenics may indicate the genetic predisposition to schizophrenia. The presence of subclinical negative symptomatology may reflect less than complete expression of the schizophrenia genotype. Results from our ongoing family study have provided some confirmation of this hypothesis (Tsuang et al. 1991). In preliminary analyses, over 20% of the relatives of schizophrenics had negative symptoms (specifically, flatness of affect and avolition) compared to only 8% of the relatives of depressed probands. The difference is significant, even when family members with diagnosable Axis-I and Axis-II psychiatric disorders were removed from the analysis. We are collecting additional negative symptom data to help us define an optimal cutpoint for "affected" in linkage analyses.

Pleiotropy

In contrast to genetic heterogeneity is phenotypic heterogeneity, which suggests that a given pathogenic genotype can be expressed as one of several phenotypes. It is not uncommon for a single gene to have different phenotypic effects in different individuals; this is known as pleiotropy. Mathematical modeling and linkage research should closely attend to the specification of phenotypes that may be genetically related to the disorder. Ideally, phenotype specification should combine traditional diagnostic distinctions with the assessment of potential biological markers.

An innovative approach to the modeling of single gene disorders was suggested by Matthysse (1985) and later applied to schizophrenia (Matthysse et al. 1986; Holzman et al. 1988). The model assumes the existence of a latent trait that is not directly observable and, depending upon its site of involvement in the brain, can cause schizophrenia or other phenotypic manifestations. It is hypothesized that the latent trait displays mendelian transmission while the observable traits (e.g., schizophrenia, schizotypal PD, etc.) do not necessarily conform to such a genetic pattern. Matthysse et al. (1986) and Holzman and colleagues (1988) have suggested that schizophrenia and smooth pursuit eye movement (SPEM) dysfunctions may be transmitted as independent phenotypic manifestations of a single latent trait (i.e., a gene). Applying the latent structure model to two different samples, the authors obtained strikingly similar and consistent results. Their findings suggest that SPEM dysfunctions and schizophrenia may be considered expressions of a single underlying trait that is transmitted by an autosomal dominant gene. Their results are not definitive because the latent trait model may not adequately account for the risk to monozygotic twins and the risk to children of two schizophrenics (McGue and Gottesman 1989). Nevertheless, their work suggests that the addition of neurobiological assessments to psychiatric studies of families may be useful in finding genes that predispose to psychopathology. By evaluating nonclinical phenotypes for inclusion in our genetic analyses, we may improve our rate of correct classification.

Another issue — which does not fit our definition for a threat to diagnostic validity, but which may similarly frustrate our segregation and linkage analyses — is the possibility that some psychiatric disorders may be due to a multifactorial polygenic (MFP) process (Faraone and Tsuang 1985; Faraone et al. 1988). If the underlying mechanisms are not due to a single or small number of alleles, this will complicate our attempts to delineate the relationship between clinical phenotypes and the pathogenic genotypes. The failure to find a single major locus (SML) model that unequivocally accounts for the familial transmission of psychiatric disorders has led to the testing of MFP models. These models assume that there are many interchangeable loci, that genes at these loci have small, additive effects on the individual's predisposition, and that all individuals have some unobservable liability or predisposition. Gottesman and Shields (1967) and Hanson et al. (1977) noted five points in favor of MFP models for schizophrenia. First, like other MFP disorders, schizophrenia is found with varying severities. Second, the risk for schizophrenia is greater for persons with many schizophrenic relatives than for persons with few. Third, risk increases as a function of the severity of a schizophrenic relative's illness. Fourth, nonschizophrenic individuals from the schizophrenia spectrum can be conceptualized as having a liability close to, but not exceeding, the threshold for schizophrenia. Fifth, MFP disorders are expected to respond slowly to natural selection. Further, the MFP model is more adequate than the SML model if, as is the case for schizophrenia, concordance in monozygotic twins is substantially greater than the concordance in dizygotic twins. Thus, the twin studies of schizophrenia are consistent with the MFP model. Path analytic MFP studies support the hypothesis that schizophrenia is to a large extent a disorder with a mostly genetic multifactorial etiology. Environmental factors are important to a much lesser degree. Overall, the results suggest that the MFP model deserves serious consideration. These results cannot rule out, however, the possibility of a mixed model in which an SML component and an MFP component both exist. However, attempts to fit such a mixed model have not been able to determine the mode of transmission (Risch and Baron 1984; Vogler et al. 1990).

It has been suggested that the MFP model is a "biological null hypothesis" because of difficulties in demonstrating the association between a specific pathophysiological process and a multifactorial etiology. However, Gottesman and McGue (1990) have asserted that MFP models may be readily compatible with research into biological mechanisms. They suggest cardiovascular disease as an example of an MFP process amenable to biological study. The explication of pathophysiological processes that affect lipids has led to the identification of specific genetic loci that influence clinical outcome. Gottesman and Shields (1982) suggested that "whenever polygenic variation has been studied under laboratory conditions ...a few handleable genes have proved to mediate a large part of the genetic variance" (pp. 225-226). Plomin et al. (1991) have described how the exciting developments in molecular genetics can be integrated with the traditional approaches of quantitative genetics to study complex diseases such as psychiatric disorders. It may be possible to use phenotypic associations with quantitative trait loci (QTL) to explicate the relationship of clinical phenomena to the genotype. This approach, which has the potential to identify genes that contribute as little as 1% of the phenotypic variance (Plomin et al. 1991), may be particularly well suited to the disorders of interest to us. This newfound capability to examine the genetic determinants of quantitative variation argues against sole reliance on dichotomous clinical phenotypes. While linkage and segregation analyses require the dichotomous classification of subjects, the QTL approach is compatible with continuous variation. This supports the importance of recording and preserving the detail of clinical phenotypes, rather than relying on a relatively procrustean syndromal approach in which diverse clinical phenomena are combined under a single diagnostic rubric.

Discussion

One factor that may contribute to all of the potential threats to diagnostic validity is the loose connection between genotype and clinical phenotypes. There may be a lack of specificity between genetically influenced processes in the brain and the clinical phenomena that we observe. Many psychiatric symptoms may represent a common-final-pathway for diverse brain dysfunctions. Moreover, we may face the dilemma that the "sloppiest" symptoms to assess may be those that are most reliably produced by the pathogenic genotype. For example, symptoms such as flat affect may be a very high frequency outcome of the schizophrenia genotype; however, it may often be produced by other genetic and environmental factors. The "neatest" symptoms in terms of their "specifiability" may prove to be less reliably produced by the pathogenic genotype. The seductiveness of clinical phenotypes such as schneiderian first-rank symptoms is that they are so clearly and unambiguously described and recognized.

Given that we are going to make mistakes in our diagnostic process, which mistakes are the least costly to make and which are the most? Not all mistakes have identical consequences. For the purpose of linkage studies, false positives are more problematic. In the analysis of linkage data, the penetrance parameter, which accounts for false negatives, can be manipulated. These analyses are relatively robust to the misspecification of the penetrance parameter (Baron 1990; Martinez et al. 1989; Ott 1985). In the Amish study of bipolar disorder, varying the parameters for gene frequency and penetrance did not dramatically affect the lod score — varying the specification of the affected phenotype did lead to dramatic differences (Egeland et al. 1987; Kelsoe et al. 1989). False positives in linkage analyses result in mistaken assumptions that recombinations have taken place and substantially reduce logarithm of odds (lod) scores.

Given the problematic nature of validity in genetic research on psychiatric disorders, there are a number of practices that we may adopt. In defining who is ill, we might categorize what we believe to be different manifestations of the same gene into different levels of certainty. For example, in schizophrenia research we might use three levels, each including phenotypes known to be increased among relatives of schizophrenics. The first level could include schizophrenia and schizoaffective disorder, depressed. Since these are severe disorders that usually require hospitalization, we are likely to have records available when making diagnoses. The increased amount of information available for these disorders and the severity of their clinical presentation is likely to reduce the false-positive rate compared to definitions at levels two and three. The second level could include schizotypal PD and psychotic disorder not otherwise specified (NOS) in the absence of substance abuse or dependence. Although psychosis NOS is similar in severity to the level one diagnoses, we suggest level two status because the diagnosis is often made when there is inadequate information for making a more specific diagnosis. We expect this to increase the false-positive rate beyond that expected for level one diagnoses. Cases of psychosis NOS that occur in the context of a continuous history of substance abuse or dependence should be defined as phenotype unknown. The third level might include those who exceed some specified cutpoint on a standardized measure of negative symptoms or demonstrate some psychophysiological or neuropsychological deficit associated with schizophrenia such as impaired performance on the Continuous Performance Task or eye movement dysfunction. Individuals with other psychotic diagnoses and those with other Axis-I disorders that have required psychiatric hospitalization should be defined as "phenotype unknown" for linkage analyses.

We recommend three methods to minimize the false-positive rate. The first line of defense against false positives is the information we collect. Structured psychiatric interviews should be specifically designed for genetic studies. This, coupled with rigorous interviewer training and comprehensive diagnostic procedures, will increase both the sensitivity and specificity of final diagnoses. Second, the diagnostic procedure should be specifically designed to minimize the false-positive rate by using experienced diagnosticians who have received explicit training about how to reduce false positives and by using the "believe the negative rule" when two independent diagnoses of the same subject disagree. Third, a diagnostic hierarchy that allows for levels of analysis should reduce the false-positive rate.

Our research on the Harvard-Brockton project indicates that multiple sources of information produce the most accurate and reliable diagnoses. Therefore, we recommend basing final diagnoses of probands and relatives on the content of an interview, a case vignette, information provided by relatives, and a complete review of the medical record by each diagnostician. Medical record reports are diagnostically useful if they provide either descriptions of observable behavior or a record of the patient's self-report of behavior or mental status.

One should be certain that all criteria are positive before assigning a diagnosis. For example, our experience is that many patients with nonaffective psychoses will meet all criteria for schizophrenia except one (e.g., for criterion A the patient may have nonbizarre delusions with auditory hallucinations that cannot be characterized as prominent). Such patients should be diagnosed as psychosis NOS, not schizophrenia. We also suggest the utility of informant information in determining the existence of manic episodes. It is often the case that bipolar patients deny a history of manic episodes either because they do not experience mania as abnormal, or for other reasons. In such cases, information from relatives and the medical record may be especially useful.

In making PD diagnoses, we recommend adhering to the DSM-III-R convention that PDs be diagnosed only if the syndrome is characteristic of the subject's functioning since early adulthood and has been a source of significant impairment or subjective distress. This is supported by the work of Rice (Rice et al. 1986, 1987a) that suggests that higher degrees of severity (e.g., psychiatric hospitalization,

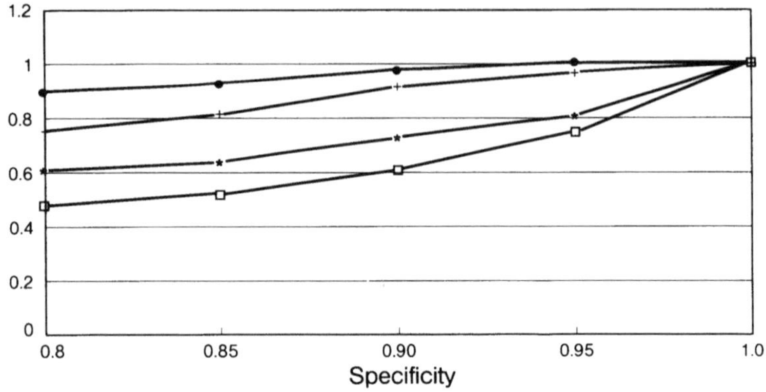

Fig. 2. True-positive rate. Believe negative, —●— ; believe third diagnosis, —+— ; one idagnosis, —*— ; believe positive, —□—

number of symptoms, need for medication) are predictive of an increased likelihood of being a true case.

For the case of disagreement between the two diagnosticians, there are four possible courses of action: (1) have them meet to decide on a consensus diagnosis; (2) believe the positive diagnosis; (3) believe the negative diagnosis; and (4) believe whichever diagnosis is supported by a third diagnostician. We do not recommend the first approach because the outcome of this procedure may be unduly influenced by idiosyncratic characteristics of the diagnosticians and may therefore not be easily repeatable by other investigators. The choice between options 2, 3, and 4 can be guided by research in epidemiology and biostatistics regarding the statistical characteristics of repeated diagnostic tests (Ekstrom et al. 1990; Goldberg and Wittes 1978; Hui and Walter 1980; McClish and Quade 1985; Nissen-Meyer 1964; Politser 1982). This literature indicates that the sensitivity and specificity of diagnosis will vary depending upon the strategy utilized for combining information from repeated tests.

Believing the positive diagnosis increases the sensitivity but decreases the specificity. Conversely, believing the negative diagnosis increases the specificity but decreases the sensitivity. Believing a third diagnostician is worth considering because it increases the sensitivity and also increases the specificity (McClish and Quade 1985). However, for the purposes of linkage analysis, the best diagnostic design is one that minimizes the proportion of false positives. By assigning values to the parameters of sensitivity, specificity, and prevalence it is possible to compute the resulting true-positive rate and the proportion of subjects diagnosed. Figure 2 compares these designs assuming the sensitivity of a single diagnostician for the illness is .85 and the prevalence of the illness is .25. As expected, the fraction of false positives among all individuals given the diagnosis decreases as the specificity increases for all designs. As Fig. 2 indicates, believing the negative diagnosis provides the best protection from false positives over a range of specificities. Believing the third diagnostician produces approximately 10% more false positives than the "believe the negative rule" if the diagnosis does not have a very high specificity. Believing the positive or relying on only a single diagnostician produces high false-positive rates. We found similar results for sensitivities ranging from .7 to .99 and prevalences ranging from .1 to .4. Of course, reducing the false-positive rate may be to no avail if it dramatically reduces the total number of individuals classified as ill.

In Figure 3 we compare each design on the proportion of all subjects given a positive diagnosis, assuming the sensitivity and prevalence used in Fig. 2. Believing the negative produces the fewest positive diagnoses, but its rate of case identification does not vary dramatically from the true prevalence of the illness. We find that similar results hold for sensitivities ranging from .7 to .99 and prevalences ranging from .1 to .4. Thus, we recommend resolving diagnostic disagreements by believing the negative diagnosis when two diagnosticians disagree; this method has a low false-positive rate and is less expensive than using a third diagnostician.

The "believe the negative rule" has several procedural implementations, as follows: (1) If the two diagnosticians disagree on diagnosis but agree on level of the hierarchy, include the subject as ill for that level of analysis. For example, if the discrepant diagnoses are schizophrenia and schizoaffective disorder, depressed, then the subject should be counted as ill for all linkage analyses. (2) If the two diagnoses are at different levels of the hierarchy, then the subject should only be included as ill in analyses at the lower level. For example, if the discrepant diagnoses are schizophrenia and psychosis NOS in the absence of substance abuse or dependence, then the subject should be included at level two of the hierarchy (i.e., believe the psychosis NOS diagnosis). (3) If one diagnosis is included at any level of the hierarchy but the other is not, believe the latter and count the subject as phenotype unknown in all linkage analyses.

In preparing this paper we have attempted to integrate the growing methodological literature pertaining to the linkage analysis of complex diseases (Diehl and Kendler 1989; Garver et al. 1989; Gurling et al. 1989; Ott 1990; Risch 1990; Lander 1988; Merikangas et al. 1989). Although there is not complete agreement on all research design issues, there is a consensus on the following methodological

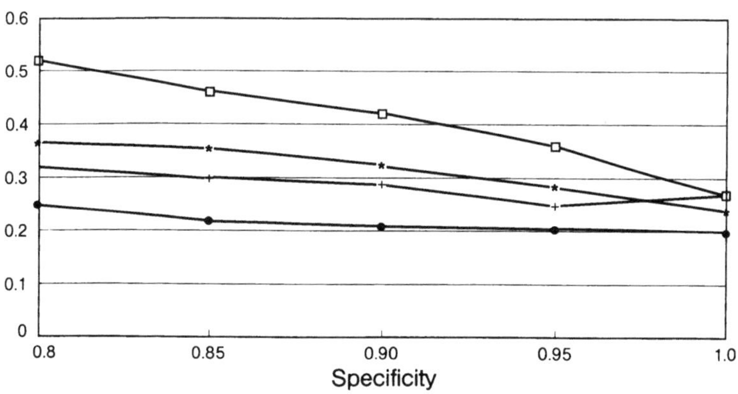

Fig. 3. Proportion of subjects diagnosed. Believe negative, —●— ; believe third diagnosis, —+—; one diagnosis, —*— ; believe positive, —□—

principles: (1) use standardized diagnostic criteria; (2) define diagnoses that will be included as affected cases before data collection; (3) use assessment and diagnostic procedures that minimize false-positive diagnoses; (4) ascertain pedigrees and collect data in a manner that can be reproduced by other investigators; (5) collect detailed clinical and demographic data to allow comparisons with other samples; (6) maintain complete blindness between the psychiatric diagnoses and marker statuses of all subjects; (7) implement procedures to facilitate the follow-up of pedigree members; (8) implement procedures to minimize laboratory errors; (9) use a threshold of statistical significance that takes into account the data analytic issues unique to complex nonmendelian disorders; (10) allow other investigators access to complete pedigree and clinical information relevant to any publications of linkage results.

As we have reviewed elsewhere (Tsuang and Faraone 1990; Faraone et al. 1990; Tsuang et al. 1991), the history of psychiatric linkage research is marred by the inability of researchers to replicate what initially appear to be impressive linkage findings, e.g., linkages to chromosomes 11 (Egeland et al. 1987) and 6 (Turner and King 1983) for bipolar disorder and chromosome 5 (Sherrington et al. 1988) for schizophrenia. Although one can appeal to genetic heterogeneity as an explanation for these results, the field is unlikely to progress if linkage findings are not independently replicated. It may not be possible to definitively specify all of the phenotypic manifestations of the genotypes for psychiatric disorders until we have identified the relevant genotypes. However, in the meantime, we must be rigorous but flexible in our assessment of potentially relevant clinical phenomena.

References

Akiskal HS, Djenderedjian AH, Rosenthal R, Khani M (1977) Cyclothymic disorder: validating criteria for inclusion in bipolar affective group. Am J Psychiatry 134:1227-1233

Andreasen NC, Rice J, Endicott J, Reich T, Coryell W (1986) The family history approach to diagnosis. Arch Gen Psychiatry 43:421-429

Baron M, Gruen R, Rainer JD, Kane J, Asnis L, Lord S (1985) A family study of schizophrenic and normal control probands: implications for the spectrum concept of schizophrenia. Am J Psychiatry 145:447-455

Baron M, Risch N, Hamburger R, Mandel B, Kushner S, Newman M, Drumer D, Belmaker RH (1987) Genetic linkage between X-chromosome markers and bipolar affective illness. Nature 325:289-292

Baron M, Hamburger R, Sandkuyl LA, Risch N, Mandel B, Endicott J, Belmaker RH, Ott J (1990) The impact of phenotypic variation on genetic analysis: application to X-linkage in manic-depressive illness. Acta Psychiatr Scand 82:196-203

Blacker D, Tsuang MT (1991) Uncharted boundaries of bipolar disorder. CME syllabus & proceedings summary, American Psychiatric Association, New Orleans

Cassano GB, Akiskal HS, Musetti L, Perugi G, Soriani A, Mignani V (1989) Psychopathology, temperament, and past course in primary major depressions. 2. Toward a redefinition of bipolarity with a new semistructured interview for depression. Psychopathology 22:278-288

Coryell WH, Zimmerman M (1989) Personality disorder in the families of depressed, schizophrenic, and never-ill probands. Am J Psychiatry 146:496-502

Detera-Wadleigh SD, Berrettini WH, Goldin LR, Boorman D, Anderson S, Gershon ES (1987) Close linkage of c-Harvey-ras-1 and insulin to affective disorder is ruled out in 3 North American pedigrees. Nature 325:806-808

Diehl SR, Kendler KS (1989) Strategies for linkage studies of schizophrenia: pedigrees, DNA markers, and statistical analyses. Schizophr Bull 15:403-419

Egeland JA, Hostetter AM (1983) Amish study I: affective disorders among the Amish, 1976-1980. Am J Psychiatry 140:56-61

Egeland JA, Gerhard DS, Pauls DL, Sussex JN, Kidd KK, Allen CR, Hostetter AM, Houseman DE (1987) Bipolar affective disorders linked to DNA markers on chromosome 11. Nature 325:783-787

Ekstrom D, Quade D, Golden RN (1990) Statistical analysis of repeated measures in psychiatric research. Arch Gen Psychiatry 47:770-772

Endicott J, Nee J, Andreasen NC, Clayton P, Keller W, Coryell W (1985) Bipolar II, combine or keep separate? J Affect Dis 8:17-28

Faraone SV, Tsuang MT (1985) Quantitative models of the genetic transmission of schizophrenia. Psychol Bull 98:41-66

Faraone SV, Lyons, MJ, Tsuang MT (1988) Schizophrenia: mathematical models of genetic transmission. In: Tsuang MT, Simpson JC (eds) Handbook of schizophrenia, vol 3. Elsevier, Amsterdam

Faraone SV, Kremen WS, Tsuang MT (1990) The genetic transmission of major affective disorders: mathematical models and linkage analyses. Psychol Bull 108:109-127

Garver DL, Reich T, Isenberg KE, Cloninger CR (1989) Schizophrenia and the question of genetic heterogeneity. Schizophr Bull 15:421-430

Goldberg JD, Wittes JT (1978) The estimation of false negatives in medical screening. Biometrics 34:77-86

Goodwin FK, Jamison KR (1990) Manic-depressive illness. Oxford University Press, New York

Gottesman II, McGue M (1990) Mixed and mixed-up models for the transmission of schizophrenia. In: Cichetti D, Grove W (eds) Thinking clearly about psychology: essays in honor of Paul E. Meehl. University of Minnesota Press, Minneapolis

Gottesman II, Shields J (1967) A polygenic theory of schizophrenia. Proc Natl Acad Sci USA 58:199-205

Gottesman II, Shields J (1982) Schizophrenia: the epigenetic puzzle. Cambridge University Press, Cambridge

Gurling HMD, Sherrington RP, Brynjolfsson J, Read T, Curtis D, Mankoo BJ, Potter M, Petursson H (1989) Recent and future molecular genetic research into schizophrenia. Schizophr Bull 15;373-382

Hanson DR, Gottesman II, Meehl P (1977) Genetic theories and the validation of psychiatric diagnoses: implications for the study of children of schizophrenics. J Abn Psychol 86:575-588

Hodgkinson S, Sherrington R, Gurling H, Marchbanks R, Reeders S, Mallet J, McInnis M, Petursson H, Brynjolfsson J (1987) Molecular genetic evidence for heterogeneity in manic depression. Nature 325: 805-806

Holzman PS, Matthysse S (1991) Review: the genetics of schizophrenia. Psychol Sci 1:279-286

Holzman PS, Kringlen E, Matthysse S, Flanagan SD, Lipton RB, Cramer G, Levin S, Lange K, Levy DL (1988) A single dominant gene can account for eye tracking dysfunctions and schizophrenia in offspring of discordant twins. Arch Gen Psychiatry 45:641-647

Hui SL, Walter SD (1980) Estimating the error rates of diagnostic tests. Biometrics 36:167-171

Kelsoe JR, Ginns EI, Egeland JA, Gerhard DS, Goldstein AM, Bale SJ, Pauls DL, Long RT, Kidd KK, Conte G, Houseman DE, Paul SM (1989) Reevaluation of the linkage relationship between chromosome 11p loci and the gene for bipolar affective disorder in the old order Amish. Nature 342:238-243

Kendler KS, Gruenberg AM, Strauss JS (1981) An independent analysis of the Copenhagen sample of the Danish adoption study of schizophrenia: II. The relationship between schizotypal personality disorder and schizophrenia. Arch Gen Psychiatry 38:982-984

Kendler KS, Gruenberg AM, Tsuang MT (1983) The specificity of DSM-III schizotypal symptoms. Abstracts of 135th Meeting of American Psychiatric Association. APA, Washington, DC

Kendler KS, Masterson CC, Ungaro R, Davis KL (1984) A family history study of schizophrenia-related personality disorders. Am J Psychiatry 141:424-427

Kendler KS, Masterson CC, Davis KL (1985) Psychiatric illness in first-degree relatives of patients with paranoid psychosis, schizophrenia, and medical illness. Br J Psychiatry 147:524

Kety SS (1988) Schizophrenic illness in the families of schizophrenic adoptees: findings from the Danish national sample. Schizophr Bull 14:217-222

Kety SS, Rosenthal D, Wender PH, Schulsinger F (1968) The types and prevalence of mental illness in the biological and adoptive families of adopted schizophrenics. J Psychiatr Res 1[Suppl]:345-362

Kety SS, Rosenthal D, Wender PH, Schulsinger F, Jacobsen B (1975) Mental illness in the biological and adoptive families of adopted individuals who have become schizophrenic. In: Fieve RR, Rosenthal D, Brill H (eds) Genetic research in psychiatry. Johns Hopkins University Press, Baltimore

Lander ES (1988) Mapping complex genetic traits in humans. In: KE Davies (ed) Genome analysis: a practical approach. IRL, Washington

Martinez M, Khlat M, Leboyer M, Clerget-Darpoux F (1989) Performance of linkage analysis under misclassification error when the genetic model is unknown. Genet Epidemiol 6:253-258

Matthysse S (1985) Genetic latent structure analysis. In: Sakai T, Tsuboi T (eds) Genetic aspects of human behavior. Igaku-Shoin, Tokyo, pp 103-111

Matthysse S, Holzman PS, Lange K (1986) The genetic transmission of schizophrenia: application of Mendelian latent structure analysis to eye tracking dysfunctions in schizophrenia and affective disorder. J Psychiatr Res 20:57-65

McClish D, Quade D (1985) Improving estimates of prevalence by repeated testing. Biometrics 41:81-89

McGue M, Gottesman II (1989) Genetic linkage in schizophrenia: perspectives from genetic epidemiology. Schizophr Bull 15:453-464

Merikangas KR, Spence A, Kupfer DJ (1989) Linkage studies of bipolar disorder: methodological and analytic issues. Arch Gen Psychiatry 46:1137-1141

Morton NE (1956) The detection and estimation of linkage between the genes for elliptocytosis and the Rh blood type. Am J Hum Genet 8:80-96

Nissen-Meyer S (1964) Evaluation of screening tests in medical diagnosis. Biometrics 20:730-755

Ott J (1985) Analysis of human genetic linkage. Johns Hopkins University Press, Baltimore

Ott J (1990) Cutting a Gordian knot in the linkage analysis of complex human traits. Am J Hum Genet 46: 219-221

Plomin R, McClearn GE, Owen M, McGuffin (1991) Quantitative genetics, molecular genetics, and dimensions of normal variation. Behav Brain Sci (in press)

Politser P (1982) Reliability, decision rules, and the value of repeated tests. Medical Decis Making 2(1):47-69

Rice JP, McDonald-Scott P, Endicott J, Coryell W, Grove WM, Keller MB, Altis D (1986) The stability of diagnosis with an application to bipolar II disorder. Psychiatr Res 19:285-296

Rice JP, Endicott J, Knesevich MA, Rochberg N (1987a) The estimation of diagnostic sensitivity using stability data: an application to major depressive disorder. J Psychiatr Res 21(4):337-345

Rice J, Reich T, Andreasen NC, Endicott J, Van Eerdewegh M, Fishman R, Hirschfeld RMA, Klerman GL (1987b) The familial transmission of bipolar illness. Arch Gen Psychiatry 44:441-447

Risch N (1990) Genetic linkage and complex diseases, with special reference to psychiatric disorders. Genet Epidemiol 7:3-7

Risch N, Baron M (1984) Segregation analysis of schizophrenia and related disorders. Am J Hum Genet 36:1039-1059

Robins E, Guze SB (1970) Establishment of diagnostic validity in psychiatric illness: its application to schizophrenia. Am J Psychiatry 41959-967

Rogers KL, Winokur G (1988) The genetics of schizoaffective disorder and the schizophrenia spectrum. In: Tsuang MT, Simpson JC (eds) Handbook of schizophrenia, vol 3. Elsevier, Amsterdam

Sherrington R, Brynjolfsson J, Petursson H, Potter M, Dudleston K, Barraclough B, Wasmuth J, Dobbs M, Gurling H (1988) Localization of a susceptibility locus for schizophrenia on chromosome 5. Nature 336:164-167

Siever LJ, Gunderson JG (1983) The search for a schizotypal personality: historical origins and current status. Comp Psychiatry 24:199-212

Squires-Wheeler E, Skodol AE, Bassett A, Erlenmeyer-Kimling L (1989) DSM-III-R schizotypal personality traits in offspring of schizophrenic disorder, affective disorder, and normal control parents. J Psychiatr Res 23(3/4):229-239

Tsuang MT, Faraone SV (1990) The genetics of mood disorders. Johns Hopkins University Press, Baltimore

Tsuang MT, Woolson RF, Winokur G, Crowe RR (1981) Stability of psychiatric diagnosis: schizophrenia and affective disorder followed up over a 30 to 40-year period. Arch Genet Psychiatry 38:535-539

Tsuang MT, Gilbertson MW, Faraone SV (1991) Genetic transmission of negative and positive symptoms in the biological relatives of schizophrenics. In: Marneros A, Tsuang MT (eds) Positive vs negative schizophrenia. (in press)

Turner WJ, King S (1983) BPD2 an autosomal dominant form of bipolar affective disorder. Biol Psychiatry 18:63-87
Vogel F, Motulsky AG (1986) Human genetics: problems and approaches (2nd edn). Springer, Berlin Heidelberg New York
Vogler GP, Gottesman II, McGue MK, Rao DC (1990) Mixed model segregation analysis of schizophrenia in the Lindelius Swedish pedigrees. Behav Genet 20:461-472
Winokur G (1985) Familial psychopathology in delusional disorder. Comp Psychiatry 26:241-248
Winokur G, Wesner R (1987) From unipolar depression to bipolar illness: 29 who changed. Acta Psychiatr Scand 76:59-63

Discussion of Presentation M. T. Tsuang et al.

Maier

You stressed avoiding false positives. If this is such a crucial point, I wonder why you did not recommend, especially for bipolar studies, using sleep characteristics as a further criterion for defining a positive case. As far as I know this is a parameter which is replicated in some family studies, also in relatives. Would you go so far as to say that true phenotypes are only those cases with affective disorders, which also show reduced REM latency.

Tsuang

If these features are proven to be stable traits for bipolar disorders, one should be encouraged to apply them on family studies as you suggested. But I am not sure whether they have been shown to be stable for genetic studies.

Mendlewicz

I do not think that the sleep data, in particular the REM latency, are that strong. The specificity of the REM latency and some other sleep measures has been questioned. Even from the recent work of David Kupffer we know it is not that specific and reliable a biological marker, thus I see no use for linkage analysis. But the question I would like to put to you concerns the definition of the phenotype. Maybe we are too naive simply to focus on illness. Because we talk about manic depression and schizophrenia, eventually we go as far as spectrum disorders. But maybe we are dealing with a more complicated problem that may be related to some trait and not an illness. I would like to have your opinion on this, because you mentioned neuropsychology. Maybe behavioral psychology has something to offer us in terms of quantification of phenotypes that are not solely based on illness definition.

Tsuang

Our group is now concentrating more and more on these specific traits. From 40 years of follow-up study of schizophrenia we learned that in the initial phase of the illness positive and negative symptoms were both present. But after some years the positive symptoms tended to become fragmented and what we saw then were more prominent negative symptoms such as flatness of affect and avolition. So we now are examining these negative symptoms in the relatives of schizophrenics from our current prospective study of major psychoses. We found that even though the relatives were not schizophrenic, the flatness of affect and avolition could be identified. So those are the characteristics we may need to concentrate on. And in the case of unipolars among the relatives of bipolars, we need to develop a scheme to rate these unipolars and to classify them into at least three subtypes: one not related to bipolars, another one probably related, and a third one definitely related to bipolars. These kinds of a more dimensional approach to diagnosis are very important, otherwise we are not going to get anywhere just by relying on categorical data.

Ackenheil

In my opinion this is a very important point. But I was surprised that you said we need specifically designed structured instruments for genetic studies. This will bring a gap between clinical psychiatry and genetics and so I am not sure if this is the right way to have specifically designed instruments for genetic studies.

Tsuang

Yes, we raised that in the NIMH Genetics Initiative research group. In order to identify specific features in the relatives with spectrum disorders and other subclinical forms, and some nonspecific personality traits, we need an instrument designed specifically for these purposes.

Cloninger

Just a comment on the issue of validity. You have listed several different criteria for validity. I think here we have to emphasize the familial validity and specifically that probably requires us to do studies, to look at the combination of signs and traits, and so to maximize genetic relationship. And that requires family studies, but also reliability studies, where we can look at signs and symptoms that predict differences between monozygotic and dizygotic twins or adoptees. That is unfortunately often missed.

Maier

I would like to come back to a point you made in the discussion. If affective flattening is such an important feature for genetic studies, I think you come into conflict with your criterion, because affective flattening will be one of the first symptoms on which two diagnosticians can disagree. They will probably not disagree with regard to first-rank schneiderian symptoms, but with regard to that criterion. Now, if this is so relevant, your criterion of agreement between the diagnosticians is not very convincing in picking out peculiar, probably true-positive cases.

Tsuang

You may know, blunting affect was dropped from the Feighner cirteria for schizophrenia when they were applied to the Iowa 500 study because of the concern with low reliability. Also, the flatness of affect may be confused with the effect of neuroleptics or with depression. Nevertheless, we have found significant flatness of affect and avolition among the nonschizophrenic relatives of schizophrenics compared with the relatives of depressive individuals. We have subsequently performed neuropsychological tests to see whether there are any neuroanatomical correlates of the flatness of affect and avolition, for example, by looking at some evidence suggesting abnormal deficit in the temporal and prefrontal areas. Essentially I am not talking about just a description of the flatness of affect and avolition. This is only the first step; the next step is to characterize their biological basis.

Hippius
For a clinician it is a very serious predicament seeing the tremendous progress in molecular genetics and biology on the one hand, but hearing the discussion of the clinical problems on the other. Altogether, I think similar discussions took place at the time of Kraepelin and Hoche, 80 years ago. We have heard exciting results which have been correlated with diagnoses from the molecular biologists. Only in the presentation of Jacques Mallet were traits and vulnerability discussed. You stated that we should investigate the normal psychology, too. I completely agree that we have to include this in our investigations. In the future we have to think and to investigate not only on a categorical system, bur we have to investigate dimensional hypotheses too. I will not say that we have to neglect diagnoses completely, but the diagnosis must be as good as possible by DSM-IV and V or whatever we have in the future. Besides that, for genetic research we need not only a framework of diagnoses, we need other frameworks such as normal personality traits, vulnerability, types of course of illness, and so on. Only if we bear this in mind will we have the sympathy of our molecular genetic colleagues in the future. If we do not, then I predict that after several years of intensive research, the molecular geneticists will turn to other fields.

Tsuang
I could not agree more. That is the area which we are pursuing. Yesterday we talked about the collection of brains from patients. We should also tress the collection of normal brains to be used for control data in these studies. Otherwise the brain tissue which we are studying could be difficult to interpret. Without control brains we do not know whether the results from the studies of brains from patients are artifacts or real findings.

Crow
I recollect a famous paper in the *Archives* entitled "Schizoaffective Disorder, Dead or Alive" by M.T. Tsuang. If you had given the answer to that question, it sounds like the answer would be "dead". Schizomanic disorder goes to affective disorder and schizodepressive disorder goes to schizophrenia. Is that right?

Tsuang
Taken in general that is true, but I do not want to close the door. So, in my recent paper based on the Eliot Slater lecture, published in the *British Journal of Psychiatry*, I

presented the morbidity risks of major psychoses among the relatives of individuals with schizoaffective disorders. In that paper, I used a very strict definition of schizoaffective disorder by excluding patients with typical manic, depressive, and schizophrenic features. In general it is now well accepted among most researchers that the schizoaffective manic type is bipolar and the depressed type is more heterogeneous. In my view there may be another type which is undifferentiated. If one uses a very narrow concept of schizophrenia and affective disorders, most of the psychoses are schizoaffective. If one uses a broad concept of schizophrenia or affective disorders, the schizoaffectives should be very rare. One of the reasons I am interested in schizoaffective disorder is not merely for the study of the disorder per se, but to study the boundary of schizophrenia and the boundary of affective disorders.

Gurling
This goes back to the point Dr. Hippius made concerning defining psychiatric phenotypes as a quantitative variable. It also relates to the issue of phenocopies and false positives. With current linkage analysis you can define phenotypes as a quantitative variable by using the linkage program in particular. But, one of the problems is that you also have to put in a preference for the unaffected homozygotes, the so-called phenocopy rate, which may reflect the prevalence of the disorders. So you can correct for false positives to some extent. We have created a quantitative variable, published in the *American Journal of Human Genetics,* taking into account the certainity of a true case as well as the population prevalence, and calculated straightforward lod scores and compared the results with the existing data. It turns out that it works quite well. There are enough methods to take into account quantitative variables and false positives.

Tsuang
I would like to emphasize that, as a first step, traits of interest should be studied among the relatives of the patients with typical schizophrenia and affective disorder, with the additional step of looking for the biological basis for such traits. If there is a biological basis for a trait, it will have important scientific value. Our group is currently trying to quantify various traits, for example, by simultaneously measuring neuropsychological dysfunction and evoked potentials and performing MRI in those relatives with flatness of affect and avolition. In this manner we hope to fully characterize informative traits and relate them to issues of nosology, epidemiology, and genetics. Just a description of a specific trait is important, but the ultimate goal is to find a biological basis for the trait.

Reich
The theory of correlated characters is quite extensive. It is important but not necessary to either use categorical or quantitative measurements; both combined

vastly amplify the power of detectability of underlying mechanisms. Steve Molden will present a poster at the upcoming meeting at Washington showing the improvement in detectability of a locus, when one has in addition to diagnosis correlated quanti-tative measurements. The power increase with respect to quantitative measure has been studied in animal genetics for long time, and so quantitative measures are used for genetic selection. The development of programs will certainly be in a position to use it the same time and the same data base when looking for latent genotypes that are relevant to mental disorders, both quantitative and qualitative diagnostic variables.

Tsuang
The categorical diagnosis of typical cases should be enforced for the selection of probands. If one uses the trait for the selection of probands, one is not going to get anywhere. However, for the diagnosis of relatives, we should broaden our criteria to move into more dimensional traits; this is the principle which we apply in our epidemiological study.

Coppen
We should be reminded of the work in biological psychiatry outside genetics. I think it would be very useful and interesting to include biological variants that are now being collected in addition to the clinical diagnoses. For example, there is good evidence that serotonin is very involved in various tests which have been developed and that levels are abnormal in a proportion of patients. And we know that this has importance for the illness as we can modify the course of illness by intervening. I urge that where possible, it would be useful to apply biological variants, which are very often collected at centers other than those where genetic studies are being carried out.

Tsuang
In reviewing the biological literature, it is quite common to find that one tends to concentrate on the study of patients. Biological studies in relatives are difficult. However, if biological psychiatry and genetic psychiatry are ever going to merge, the biological traits of these families should be studied.

Biological Markers in Schizophrenia and the Affective Psychoses

D. H. R. Blackwood and W. J. Muir

The application of linkage analysis to schizophrenia and manic-depressive illness is confounded by the problem of defining the phenotypes of these conditions which are classified almost entirely by signs and symptoms observed and elicited at clinical interview. The absence of objective biological traits shown to be associated with illness ensures a continuing debate about the validity of current classifications. It is probable that what we call schizophrenia and manic-depressive illness represent a heterogeneous group of conditions with an indistinct boundary between them. The identification of a biological trait found more frequently in a group of psychotic patients than in the general population which is independent of medication, and persists in subjects following remission from an acute episode of illness, could provide a clue to the pathogenesis of the disease. If such a trait also segregated with disease in families and was found in a proportion of unaffected relatives, it could increase the power of genetic linkage analysis by (1) leading to the selection of a more homogeneous subgroup of the disease; (2) providing a more accurate knowledge of the mode of inheritance; (3) helping to identify phenocopies; and (4) classifying asymptomatic cases, thereby increasing penetrance.

A recent example of the successful use of a biological marker in linkage studies was reported by Greenberg et al. (1988) who found evidence that juvenile myoclonic epilepsy was linked to a chromosome 6 locus. Data came from 24 families in which the proband had epilepsy. A proportion of unaffected siblings were found to have a diffuse, rather nonspecific, abnormality of the EEG. Linkage analysis carried out under the assumption that only those subjects with clinical epilepsy were "affected" yielded a logarithm of odds (lod) score below -2, excluding linkage to the chromosome 6 locus. However, when the abnormal EEG trait in unaffected relatives was considered as a subclinical marker of the disease genotype and those individuals with an abnormal EEG were included as "affected", evidence for linkage (lod > +3) was obtained. The EEG abnormality proved to be a useful trait marker even though pathogenesis was unknown and the relationship between having the trait and developing the disease was not clear.

In schizophrenia and manic-depressive illness, many atypical physiological, neuroanatomical, pharmacological and neuropsychological findings have been reported (Baron 1986). As outlined by Rieder and Gershon (1978), a trait which is to be considered a subclinical marker for the disease genotype should be found more commonly in the diseased population and in relatives of probands should be genetically transmitted and be independent of clinical state. In this context specificity of the trait to one particular disease and a knowledge of relationship between the expression of the trait and the clinical expression of the disease may actually be of little importance (Lander 1988). One of the earliest reports of a biological marker for the psychoses was made by Diefendorf and Dodge (1908) who showed impaired pursuit eye movements in patients with "dementia praecox". Since then a considerable number of biological traits possibly related to genetic susceptibility to psychoses have been identified. For example, biochemical traits reflecting brain monoamine metabolism as monitored by enzyme and metabolite assays in peripheral blood and cerebrospinal fluid have been examined. Amongst the best studied are platelet monoamine oxidase levels (Wyatt et al. 1973) and spiperone binding to lymphocytes (Le Fur 1983; Bondy et al. 1985) (reviewed by Goldin et al. 1987). Extensive research has also been made into physiological changes associated with psychoses such as disorders in the smooth pursuit eye movement system and disturbances in electroencephalographic event-related potentials.

Smooth Pursuit Eye Movement in Schizophrenia and Manic Depression

When a human volunteer watches a moving target and his or her eye movements are followed by recording the electrooculogram or by means of infrared light reflected by the iris and sclera onto sensors attached to eye frames, two kinds of eye movement are observed. When the target object is moving slowly (e.g. sinusoidally at 0.4 Hz), "smooth pursuit" movements of the eye enable the subject to maintain a stable image on the retina. Other types of eye movement include saccades, which are rapid step-like movements which enable the eye to quickly fix on a target. These different types of eye movements are under quite separate neuronal control. Holzman et al. (1973), studying schizophrenic patients, reported impaired smooth pursuit eye movement consisting of an inefficiency of the smooth pursuit system, resulting in an excess of saccadic intrusions and bursts of saccadic tracking with normal latencies and velocities of the saccades themselves (Iacono et al. 1981; Levin et al. 1981, 1982; Mather and Puchat 1982). Between 50% and 80% of the schizophrenic population as opposed to 8% of normal controls had eye movement dysfunction as detected by visual inspection of the electrooculographic tracings. Neuroleptic medication in schizophrenics did not affect smooth pursuit movement (Shagass et al. 1974; Holzman and Levy 1977; Iacono et al. 1981) and the abnormality persisted with remission of psychotic symptoms (Levy et al. 1983). The use of eye tracking dysfunction as a biological marker for schizophrenia was suggested by the results of family studies showing that around 50% of first-degree relatives of schizophrenics had eye tracking dysfunction as opposed to only 8% of control subjects (Holzman et al. 1974, 1984). The pairwise concordance rate for eye tracking disorder in monozygotic twins discordant for the clinical diagnosis of schizophrenia was higher than that found in dizygotic twins (Holzman et al. 1977, 1978, 1980), suggesting that genetic factors contributed to the variability of smooth pursuit eye movements. An important observation in these twin studies was that pursuit eye movements were sometimes quite normal in a schizophrenic twin but abnormal in the clinically unaffected co-twin. Similarly, family studies sometimes revealed that a clinically unaffected parent had abnormal eye movements but their schizophrenic offspring had normal eye movements (Holzman 1987). To account for these findings, it was proposed that an underlying "latent trait", a presumed single gene defect, could phenotypically present itself as schizophrenia or eye movement disorder, or as a combination of both (Matthysse et al. 1986; Matthysse and Holzman 1987; Holzman et al. 1988; Holzman 1989). Clinical symptoms of schizophrenia and eye movement disorder would thus be viewed as different phenotypic expressions of a single gene. This latent trait model was successfully fitted to family data (Holzman et al. 1988) but its validity has been challenged (McGue and Gottesman 1989) on the grounds that it fails to predict accurately the risk of schizophrenia among the relatives of schizophrenic probands. While there is now convincing evidence for a familial association between eye movement disorder and schizophrenia, there is uncertainty about the specificity of the disorder within the group of functional psychoses. Eye tracking disorder was measured in patients with affective disorders (Shagass et al. 1974; Klein et al. 1976; Salzman et al. 1978; Lipton et al. 1980; Levin et al. 1981; Iacono et al. 1982). In contrast to the familial tendency of the disorder in schizophrenia, the first-degree relatives of patients with bipolar illness do not seem to show an increased rate of dysfunction (Levy et al. 1983; Holzman et al. 1984). Lithium therapy for bipolar illness may also impair eye tracking performance (Levy et al. 1985; Holzman et al. 1991). A recent study by Muir et al. (1991a) found that smooth pursuit disorder does distinguish between patients with schizophrenia and manic-depressive illness. Smooth pursuit eye movements were recorded using the electro-oculogram in 49 subjects with bipolar disorder, 19 with major depressive (unipolar) disorder and 61 with definite schizophrenia and compared with 145 normal controls. Smooth pursuit was significantly more disordered in the schizophrenic group than in bipolars, major depressed or controls, and the finding was independent of the effect of neuroleptics, tricyclic antidepressants or lithium therapy. Evidence that eye tracking impairment may also be a useful marker for schizophrenia spectrum disorder comes in a report by Siever et al. (1990) on 26 patients with schizotypal personality disorder who had eye tracking dysfunction indistinguishable from a group of 64 schizophrenic patients.

Only a relatively small number of studies have attempted to relate eye movement dysfunction to neuropsychological or neuroanatomical changes in schizophrenia. Blackwood et al. (1991a) reported a significant correlation between eye movement dysfunction and ventricular size measured by magnetic resonance imaging (MRI) scanning in a group of 31 schizophrenics, but Katsanis et al. (1991) found no association between eye tracking and ventricular size measured by computed tomography (CT) scanning in a group of 36 schizophrenic patients. Levin (1984a,b) proposed that the eye movement disorder of schizophrenics was associated with frontal lobe dysfunction and there is neuropsychological support for this view from two studies by Bartfai et al. (1985, 1989). Schizophrenic patients with eye movement dysfunction showed poorer performance on neuropsychological tests thought to be sensitive to frontal and frontoparietal impairment. Similarly, clinically well volunteers with eye tracking dysfunction showed similar, though less pervasive, impairment on these tests. Coursey et al. (1989) showed a relationship between poor tracking and general intellectual decrements in a group of college students.

Abnormal Event-Related Potentials in Schizophrenia and Manic-Depressive Illness

Event-related potentials, as the name suggests, are the very small voltages generated in brain structures in response to specific events or stimuli. For example, during the first 10 ms following an auditory click stimulus, event-related potentials recorded from the human scalp include peaks resulting from neuronal activity in the acoustic nerve, cochlear nucleus, superior olive, lateral lemniscus and inferior colliculus. Similar early event-related potentials are generated in the visual pathways following a flash of light and in the spinal cord and sensory afferent pathways of the brainstem following a peripheral somatosensory stimulus. These early "exogenous" potentials are extremely useful clinically for detecting disorders of sensory endorgans and of the brainstem. However, of greater interest to psychiatry are the later so-called "endogenous potentials" generated after about 50 ms which reflect different aspects of information processing by the subject at a higher cortical level. The latency and amplitude of these late cognitive potentials may depend on how motivated the subject is and on their cognitive, affective or motor response to a particular stimulus. The P300 response, a positive wave some 300 ms following a stimulus, is generated during the performance of a task which requires the resolution of uncertainty. During an "oddball" task, the subject listens to two different stimuli (e.g. a high-pitched and a low-pitched tone), one of which occurs relatively infrequently and is designated as a target. A P300 response may be generated following auditory, visual or somatosensory stimuli when a subject recognises and responds to a rare target signal embedded in a series of more frequent unattended signals. A very large literature on P300 (Regan 1989) has addressed, but not yet resolved, the question of the cognitive processes contributing to the generation of this P300 waveform. P300 latency increases with the difficulty of the task (e.g. having to distinguish between three tones rather than just two), and the amplitude of P300 increases with the unexpectedness of the target tone. The P300 is thought to be a complex mixture of at least three separate waveforms. These reflect different aspects of information processing in the recognition, retrieval from memory and judgements about the significance of the target signal. P300 seems to reflect mental processes that allow us to anticipate significant events in the environment and to react to unexpected changes. It is therefore not surprising that abnormalities in these cognitive potentials should have been found in psychiatric disorders, including schizophrenia and manic-depressive illness. Event-related potentials are noninvasive and can be readily and reliably recorded from quite disturbed psychotic subjects. Studies have been carried out to compare event-related responses in different groups of psychiatric patients and to relate these changes to clinical, neuropsychological and brain imaging data.

P300 in Schizophrenia and Affective Disorders

Many studies have shown a reduced P300 amplitude during auditory detection tasks in schizophrenia (Roth and Cannon 1972; Levit et al. 1973; Shagass 1976; Baribeau-Braun 1983; Morstyn et al. 1983; Blackwood et al. 1987; Romani et al. 1987). A number of studies have also found that P300 latency is increased in schizophrenia (Baribeau-Braun 1983; Pfefferbaum et al. 1984; Gordon et al. 1986; Blackwood et al. 1987; Romani et al. 1987; Ebmeier et al. 1989) and in the relatives of schizophrenic subjects (Saitoh et al. 1984; Blackwood et al. 1991b). Prolonged P300 latency in schizophrenia, and to a lesser extent reduced amplitude, may be useful physiological markers for genetic studies in schizophrenia. P300 latency appears to be independent of the effects of medication and the clinical state at the time of testing (Blackwood et al. 1987). Increased P300 latency was also found in subjects with borderline/schizotypal personality disorder (Blackwood et al. 1986; Kutcher et al. 1987).

The twin studies of Lennox et al. (1945) which showed that brain electrical activity is strongly influenced by genetic factors have been well confirmed (Vogel 1970; Stassen et al. 1988). Twin studies have shown a high degree of heritability for averaged event-related potentials and, in particular, the latency of the auditory P300 response has high concordance in monozygotic pairs (Surwillo 1980; Polich and Burns 1987).

P300 latency and amplitude changes are clearly not specific to schizophrenia and may be found in several brain disorders, including the dementias. Pfefferbaum et al. (1984) compared P300 in dementia, schizophrenia, depression and controls and showed that patients with schizophrenia, like the dementia group, had an increased latency and a reduced amplitude of the auditory P300, while the depressed group did not differ from controls. P300 latency increase and amplitude reduction are also features of manic-depressive illness. In a recent study (Muir et al. 1991b) P300 was recorded from 96 subjects with schizophrenia, 99 with bipolar affective disorder, 48 with major depressive (unipolar) disorder, 32 psychiatric inpatients with a variety of nonpsychotic illnesses and 213 normal controls. The latency of the P300 component was significantly greater in the schizophrenic and bipolar subjects than in the other three groups, and this difference was stable with respect to clinical state at the time of testing and was not the result of psychotropic medications. P300 latency clearly differentiated unipolar from bipolar affective disorder but did not distinguish bipolar illness from schizophrenia.

Several studies have been designed to relate the latency and amplitude abnormalities of P300 in schizophrenia to neuropsychological or neuroanatomic changes. A bilateral decrease in P300 amplitude maximal over the left temporal area in schizophrenics has been reported (Morstyn et al. 1983; Faux et al. 1988). This amplitude reduction was significantly correlated with positive symptoms of schizophrenia and with left sylvian fissure enlargement measured from CT scans (McCarley et al. 1989). However, Pfefferbaum et al. (1989) found no topographic differences in P300 amplitude between control and schizophrenic subjects. Romani et al. (1987) performed CT scans and recorded P300 on 20 schizophrenic subjects but found no consistent association between ventricular enlargement and P300 abnormalities. Using MRI to scan 31 schizophrenic subjects and 33 volunteer controls, Blackwood et al. (1991a) found a modestly significant positive correlation between ventricular brain ratio and P300 latency. These schizophrenic subjects with abnormal P300 latency also showed a significant reduction in the area of the cingulate cortex bilaterally and P300 latency correlated with changes in the sizes of the amygdala on the left and right side measured from coronal sections. In the same studies, abnormal P300 in schizophrenic subjects was associated with poor performance on a verbal fluency test and Corsi's block tapping test, thought to be sensitive to frontal and left temporal lobe impairments, respectively (Roxborough et al. 1991).

P300 and Eye Movement Disorder in Schizophrenic Families

P300 latency delay, P300 amplitude reduction and smooth pursuit eye movement disorder each have characteristics which suggest they could be useful markers in genetic studies of schizophrenia, while their role in manic-depressive illness requires some further clarification. These abnormalities, while by no means specific to schizophrenia, are found in a proportion of schizophrenic patients (in the case of P300 latency the difference between the schizophrenic and control means is approximately two standard deviations; Blackwood et al. 1987); they are under genetic control and are relatively independent of the effects of medication and clinical state. A study was therefore designed to measure auditory P300 response and smooth pursuit eye tracking in members of 20 high density schizophrenic families who had agreed to take part in linkage studies (Blackwood et al. 1991b). Details of some of these pedigrees have been published (St. Clair et al. 1989). Some 196 family members were interviewed and 45 diagnosed schizophrenic. At the time of interview, recordings were made of auditory P300 and smooth pursuit eye movements. The family members were compared to a group of 212 normal controls and 96 hospitalised schizophrenic patients. Abnormalities of P300, eye tracking or both were found in members of 17 of the 20 families. There were three families (22 subjects, including six with schizophrenia) in which no member had abnormal P300 or eye tracking and it is possible that these physiologically "normal" families form a subgroup of schizophrenic pedigrees. There were no differences in P300 or eye tracking between schizophrenic family members and the larger group of hospitalised schizophrenic controls. Both P300 and eye tracking abnormalities were also found in many nonschizophrenic relatives, amongst whom both P300 and eye tracking showed a bimodal distribution with approximately half of the nonschizophrenic relatives showing eye tracking dysfunction and/or abnormal event-related potentials. Of the 41 family members with P300 latency greater than two standard deviations above the control mean, 44% had major psychiatric illness (bipolar illness, major depression, schizoaffective, unspecified functional psychosis) but 19 subjects had no history of any psychiatric disease. Similarly, eye tracking dysfunction in these families was not restricted to schizophrenia but was found in subjects with unspecified functional psychosis, major depression, minor depression and alcoholism. Of most importance, out of 107 family members who had never been psychiatrically ill, 15 had eye tracking dysfunction. The results of this family study support the view that event-related potentials and eye movement disorder may add considerably to the power of linkage analysis in schizophrenia by helping to define, within families, the spectrum of illness, thereby improving the estimate of penetrance and reducing the risk of including phenocopies as affected individuals (Blackwood et al. 1991c). This is despite the fact that P300 latency increase and eye movement disorder are clearly not specific to the functional psychoses. For example, a reduced amplitude of P300 has been recorded in alcoholics and sons of alcoholic fathers and the P300 response may have a role in understanding the genetics of alcoholism in pedigrees (Porjasz and Begleiter 1985). It is reasonable to suppose, however, that within multiply affected schizophrenic pedigrees, P300 and eye tracking abnormalities which segregate with illness, and which are bimodally distributed amongst unaffected relatives — some of whom have symptoms, some of whom have none — reflect brain disturbance which is linked to the development of schizophrenia in the probands, although it is not clear at what point in the path from genotype to phenotype these physiological disturbances develop.

References

Baribeau-Braun J, Picton TW, Gosselin J-Y (1983) Schizophrenia: a neurophysiological evaluation of abnormal information processing. Science 219:874-876

Baron M (1986) Genetics of schizophrenia: II. Vulnerability traits and gene markers. Biol Psychiatry 21: 1189-1211

Bartfai A, Levander SE, Nyback H, Berggren B-M, Schalling D (1985) Smooth pursuit eye tracking, neuropsychological test performance and computed tomography in schizophrenia. Psychiatr Res 15:49-62

Bartfai A, Wirsen A, Levander S, Schalling D (1989) Smooth pursuit eye tracking and neuropsychological performance in healthy volunteers. Acta Psychiatr Scand 80:479-486

Blackwood DHR, St Clair DM, Kutcher SP (1986) P300 event-related potential abnormalities in borderline personality disorder. Biol Psychiatry 21:560-564

Blackwood DHR, Whalley LJ, Christie JE, Blackburn IM, St Clair DM, McInnes A (1987) Changes in auditory P3 event-related potential in schizophrenia and depression. Br J Psychiatry 150:154-160

Blackwood DHR, Young AH, McQueen JK, Martin MJ, Roxborough HM, Muir WJ, St Clair DM, Kean D (1991a) Magnetic resonance imaging in schizophrenia: altered brain morphology associated with P300 abnormalities and eye tracking dysfunction. Biol Psychiatry (in press)

Blackwood DHR, St Clair DM, Muir WJ, Duffy JC (1991b) Auditory P300 and eye tracking dysfunction in schizophrenic pedigrees. Arch Gen Psychiatry (in press)

Blackwood D, St. Clair DM, Muir W (1991c) DNA markers and biological vulnerability markers in families multiply affected with schizophrenia. Eur Arch Psychiatry Clin Neurosci 240:191-196

Bondy B, Ackenheil M, Elbers R, Frohler M (1985) Binding of H^3-spiperone to human lymphocytes: a biological marker in schizophrenia. Psychiatr Res 15:41-48

Coursey RD, Lees RW, Siever LJ (1989) The relationship between smooth pursuit eye movement impairment and psychological measures of psychopathology. Psychol Med 19:343-358

Diefendorf AR, Dodge R (1908) An experimental study of the occular reactions of the insane from photographic records. Brain 31:451-489

Ebmeier KP, Mackenzie AR, Potter DD, Salzen EA (1989) Late positive event related potentials in schizophrenia. In: Crawford J, Parker D (eds) Developments in clinical and experimental neuropsychology. Plenum, New York, pp 127-136

Faux SF, Torello MW, McCarley RN, Shenton ME, Duffy FH (1988) P300 in schizophrenia, confirmation and statistical validation of temporal region deficit in P300 topography. Biol Psychiatry 23:776-790
Goldin LR, DeLisi LE, Gershon ES (1987) Genetic aspects to the biology of schizophrenia. In: Nasrallah AH (ed) Handbook of schizophrenia, vol. II. Elsevier, Amsterdam, pp 467-487
Gordon E, Kraiuhin C, Harris A, Meares R, Honson A (1986) The differential diagnosis of dementia using P300 latency. Biol Psychiatry 21:1123-1132
Greenberg DA, Delgado-Escueta AV, Widelitz H, Sparkes RS, Treiman L, Maldonado HM, Park MS, Terasaki PI (1988) Juvenile myoclonic epilepsy (JME) may be linked to the BF and HLA loci on human chromosome 6. Am J Med Genet 31:185-192
Holzman PS (1987) Recent studies of psychophysiology in schizophrenia. Schizophr Bull 13:49-75
Holzman PS (1989) The use of eye movement dysfunctions in exploring the genetic transmission of schizophrenia. Eur Arch Psychiatry Neurol Sci 239:43-48
Holzman PS, Levy DL (1977) Smooth pursuit eye movements and functional psychoses: a review. Schizophr Bull 3:15-27
Holzman PS, Proctor LR, Hughes DW (1973) Eye tracking patterns in schizophrenia. Science 181:179-181
Holzman PS, Proctor LR, Levy DL, Yasillo NJ, Meltzer HY, Hurt SW (1974) Eye-tracking dysfunction in schizophrenic patients and their relatives. Arch Gen Psychiatry 31:143-151
Holzman PS, Kringlen E, Levy DL, Proctor LR, Haberman SJ, Yasillo NJ (1977) Abnormal pursuit eye movements in schizophrenia. Arch Gen Psychiatry 34:802-805
Holzman PS, Kringlen E, Levy DL, Proctor LR, Haberman S (1978) Smooth pursuit eye movements in twins discordant for schizophrenia. J Psychiatr Res 4:111-120
Holzman PS, Kringlen E, Levy DL, Haberman S (1980) Deviant eye tracking in twins discordant for psychosis: a replication. Arch Gen Psychiatry 37:627-631
Holzman PS, Solomon CM, Levin S, Waternaux CS (1984) Pursuit eye movement dysfunctions in schizophrenia. Arch Gen Psychiatry 41:136-139
Holzman PS, Kringlen E, Matthysse S, Flanagan SD, Lipton RB, Cramer G, Levin S, Lange K, Levy DL (1988) A single dominant gene can account for eye tracking dysfunctions and schizophrenia in offspring of discordant twins. Arch Gen Psychiatry 45:641-647
Holzman PS, O'Brian C, Waternaux C (1991) Effects of lithium treatment in eye movements. Biol Psychiatry 29:1001-1015
Iacono WG, Tuason VB, Johnson RA (1981) Dissociation of the smooth-pursuit and saccadic eye tracking in remitted schizophrenics. Arch Gen Psychiatry 38:991-996
Iacono WG, Pelequin LJ, Lumry AE, Valentine R, Tuason VB (1982) Eye tracking in patients with unipolar and bipolar affective disorders in remission. J Abnorm Psychol 91:35-44
Katsanis J, Iacono WG, Beiser M (1991) Relationship of lateral ventricular size to psychophysiological measures and short-term outcome. Psychiatr Res 37:115-129
Klein RH, Salzman LF, Jones F, Ritzler B (1976) Eye tracking in psychiatric patients and their offspring. Psychophysiology 13:186
Kutcher SP, Blackwood DHR, St. Clair DM, Gaskell DF, Muir WJ (1987) Auditory P300 in borderline personality disorder and schizophrenia. Arch Gen Psychiatry 44:645-650
Lander ES (1988) Splitting schizophrenia. Nature 336:105-106
Le Fur G, Zarifan E, Phan T, Cuche H, Flamier A, Bouchami F, Burgevin MC, Loo H, Gerard A, Uzan A (1983) H^3-spiperone binding in lymphocytes: changes in two different groups of schizophrenic patients and effect of neuroleptic treatment. Life Sci 32:249
Lennox WG, Gibbs EL, Gibbs FA (1945) The brain-wave pattern, an hereditary tract. Evidence from 74 "normal" pairs of twins. J Hered 36:233-243
Levin S (1984a) Frontal lobe dysfunctions in schizophrenia. I. Eye movement impairments. J Psychiatr Res 18:27-55
Levin S (1984b) Frontal lobe dysfunctions in schizophrenia. II. Impairments of psychological and brain functions. J Psychiatr Res 18:57-72
Levin S, Holzman PS, Rothenberg SJ, Lipton RB (1981) Saccadic eye movements in psychotic patients. Psychiatr Res 5:47-58
Levin S, Jones A, Stark L, Merrin EL, Holzman PS (1982) Identification of abnormal patterns in eye movements of schizophrenic patients. Arch Gen Psychiatry 39:1125-1130
Levit AL, Sutton S, Zubin J (1973) Evoked potential correlates of information processing in psychiatric patients. Psychol Med 3:487-494

Levy DL, Lipton RB, Holzman PS, Davis JM (1983) Eye tracking dysfunction unrelated to clinical state and treatment with haloperidol. Biol Psychiatry 18:813-819

Levy DL, Dorus E, Shaughnessy R, Yasillo NJ, Pandey GN, Janicak PG, Gibbons RD, Gavira M, Davis JM (1985) Pharmacologic evidence for specificity of pursuit dysfunction to schizophrenia: lithium carbonate associated abnormal pursuit. Arch Gen Psychiatry 42:335-341

Lipton RB, Levin S, Holzman PS (1980) Horizontal and vertical pursuit eye movements, the oculocephalic reflex and the functional psychoses. Psychiatr Res 3:193-203

Mather JA, Puchat C (1982) Motor control of schizophrenics. I. Oculomotor control of schizophrenics: a deficit in sensory processing, not strictly in motor control. J Psychiatr Res 17:343-360

Matthysse S, Holzman PS (1987) Genetic latent structure models: implications for research on schizophrenia. Psychol Med 17:271-274

Matthysse S, Holzman PS, Lange K (1986) The genetic transmission of schizophrenia: application of mendelian latent structure analysis to eye tracking dysfunctions in schizophrenia and affective disorder. J Psychiatr Res 20:57-76

McCarley RW, Faux SF, Shenton M, Le May M, Cane M, Ballinger R, Duffy FH (1989) CT abnormalities in schizophrenia. Arch Gen Psychiatry 46:698-708

McGue M, Gottesman II (1989) A single dominant gene still cannot account for the transmission of schizophrenia. Arch Gen Psychiatry 46:479

Morstyn R, Duffy FH, McCarley RW (1983) Altered P300 topography in schizophrenia. Arch Gen Psychiatry 40:729-734

Muir WJ, St. Clair DM, Blackwood DHR, Roxborough HM, Marshall I (1991a) Eye tracking dysfunction in the affective psychoses and schizophrenia. Psychol Med (in press)

Muir WJ, St. Clair DM, Blackwood DHR (1991b) Long-latency auditory event-related potentials in schizophrenia and in bipolar and unipolar affective disorder. Psychol Med (in press)

Pfefferbaum A, Wenegrat BG, Ford JM, Roth WT, Kopell BS (1984) Clinical application of the P3 component of event-related potentials. II. Dementia, depression and schizophrenia. Electroencephalogr Clin Neurophysiol 59:104-124

Pfefferbaum A, Ford JM, White PM, Roth WT (1989) P3 in schizophrenia is affected by stimulus modality, response requirements, medication status and negative symptoms. Arch Gen Psychiatry 46:1935-1044

Polich J, Burns T (1987) P300 from identical twins. Neuropsychologia 25:299-304

Porjesz B, Begleiter H (1985) Human brain electrophysiology and alcoholism. In: Tarter RD, Van Thiel D (eds) Alcohol and the brain. Plenum, New York, p 139

Regan D (1989) Human brain electrophysiology evoked potentials and evoked magnetic fields in science and medicine. Elsevier, New York

Rieder RO, Gershon ES (1978) Genetic strategies in biological psychiatry. Arch Gen Psychiatry 35:866-873

Romani A, Merello S, Gorreli , Zerri F, Grassi M, Cosi V (1987) P300 and CT scan inpatients with chronic schizophrenia. Br J Psychiatry 151:306-313

Roth WT, Cannon EH (1972) Some features of the auditory evoked response in schizophrenics. Arch Gen Psychiatry 27:466-471

Roxborough H, Muir WJ, Blackwood DHR, Walker MT, Blackburn IM (1991) Neuropsychological and P300 abnormalities in schizophrenics and their relatives. Psychol Med (in press)

St. Clair DM, Blackwood DHR, Muir WJ, Baillie D, Hubbard A, Evans HJ (1989) No linkage of chromosome 5q 11-13 markers to schizophrenia in Scottish families. Nature 339:395-309

Saitoh O, Niwa S, Hiramatsu K, Kameyama T, Rymar K, Itoh K (1984) Abnormalities in late positive components of event-related potentials may reflect a genetic predisposition to schizophrenia. Biol Psychiatry 19:293:303

Salzman LF, Klein RH, Strauss JS (1978) Pendulum eye tracking in remitted psychiatric patients. J Psychiatr Res 14:121-126

Shagass C (1976) An electrophysiological view of schizophrenia. Biol Psychiatry 11:3-30

Shagass C, Amadeo M, Overton DA (1974) Eye tracking performance in psychiatric patients. Biol Psychiatry 9:245-260

Siever LJ, Keefe R, Bernstein DP, Coccaro EF, Klar HM, Zemishlany Z, Peterson AE, Davidson M, Mahon T, Horvath T, Mohs R (1990) Eye tracking impairment in clinically identified patients with schizotypal personality disorder. Am J Psychiatry 147:740-745

Stassen HH, Lykken DT, Propping P, Bomben G (1988) Genetic determination of the human EEG. Hum Genet 80:165-176

Surwillo WN (1980) Cortical evoked potentials in monozygotic twins and unrelated subjects: comparisons of exogenous and endogenous components. Behav Genet 19:201-209
Vogel F (1970) The genetic basis of the normal human electroencephalogram. Hum Genet 19:91-114
Wyatt RJ, Murphy DL, Belmaker R, Cohen S, Donnelly CH, Pollin W (1973) Reduced monoamine oxidase

Discussion of Presentation D. Blackwood and W. J. Muir

Ott

Presumably a problem with these markers is that they simply could be consequences of the disease. You did show the distribution of some of these variables within families and you showed that it was bimodal or at least there were two components. In these distributions were the affected individuals also included?

Blackwood

The distribution of P300 latency was analyzed in the group of nonschizophrenic relatives. The schizophrenic family members were analyzed separately. The relatives' group did, however, include some subjects with major depressive disorder, bipolar illness, anxiety states, and alcoholism, but the majority had no psychiatric diagnosis.

Ott

Are there ever any two-component distributions among unaffected relatives, who would then correspond to the gene carriers and non-gene carriers?

Blackwood

We have found that 18% of unaffected relatives who had no psychiatric diagnosis had a P300 latency which could be considered abnormal (more than two standard deviations from the control mean). This group could correspond to the gene carriers.

Engel

I very much agree with the general message that you presented, which is to study biological markers in psychiatric disease. I have some problems, however, with some of the data that you presented for schizophrenia. The most problematic point is the issue of specificity of some of the results. I would say that what you have presented about P300 latency and P300 amplitude in schizophrenia could equally apply to Alzheimer's disese. I think we have to consider carefully the specificity of biological markers, otherwise we could start by searching for the genetic cause of a disease and end up by finding a gene for an EEG wave.

Blackwood

I agree that it is important to make sure that these pedigrees do not have subjects with Alzheimer's disease or other conditions apart from schizophrenia associated with prolonged P300 latency. If there were, we obviously could not use the P300 as a disease marker for schizophrenia. We can improve the specificity of P300 within families by screening for signs of dementia or a history of brain injury, epilepsy, and other neurological conditions. Presumably, any pathology affecting the brain areas responsible for P300 generation, such as the inferior parietal region, may have an influence on the P300 response. Similarly, several different pathologies may affect eye movements. Within a family we can still use these markers despite their overall lack of specificity, provided we try to exclude other neurological diseases. As far as we know, in these families there are no relatives with Alzheimer's disease.

Maier

With regard to your comment that most of these findings are of genetic origin, it is well known that ventricular brain ratio is especially high in relatives, and if there is a high loading for obstetric complications in these families, can you really be sure that the reason for this familial aggregation is genetic? It could be familial aggregation due to an environmental effect that is common to all relatives.

Blackwood

The evidence that P300 waveform is under some degree of genetic control comes from the study of normal twins. Eye tracking dysfunction has also been examined in psychotic twins. To answer your question about familial aggregation of P300 and eye tracking abnormalities being due to an environmental effect, I think this is a possibility but it does not detract from the use of these measures as markers for disease. These markers may represent disturbance at any stage on the path from genotype to phenotype.

Ackenheil

With regard to the ventricular enlargement, there are also studies on monozygotic twins in which only the affected twin has ventricular enlargement. So it mostly cannot be genetically determined.

Blackwood

In a twin study by Holzman, there was good concordance of eye tracking abnormalities in monozygotic twins discordant for psychosis. However, ventricular size was not measured in the study, so the relation between eye tracking and ventricular size in these twins is not known. I do not know of any data relating ventricular size and P300 in twins. I wonder how strict we have to be in determining the mode of inheritance of markers for these to be useful in linkage studies. For example, linkage studies in asthma were aided by the IgE response to allergens and myoclonic epilepsy by rather nonspecific EEG responses, even those these "markers" did not have a clearly understood mode of inheritance.

Merikangas

I was concerned about the specificity as well. A number of studies have shown P300 changes in alcoholics and children of alcoholics who have not yet developed the disorder. We have now also looked at migraine and anxiety disorders and also found abnormalities in the P300, although with a different pattern. So I am concerned that you might find something that is common to many different conditions, which would be expected to occur in these famlies.

Blackwood

My understanding is that alcoholics (pre-alcohol ingestion) and their relatives at high risk of alcoholism have a reduction in amplitude but no latency change of P300. I agree that many of the changes we have observed in evoked potentials are not specific to schizophrenia. However, the pattern of responses, including relative changes in amplitude and latency, could help improve specificity, as is the case in the dementias, where a different pattern of abnormalities is found in cortical and so-called "subcortical" dementias.

Merikangas

The P300 abnormality was first reported in alcoholics themselves. Several people have attempted to replicate that in offspring and we have now looked in offspring under and over 12 and we find differences in the pattern. Once people are close to maturity, we do see the increased latency and decreased amplitude. I think it has been quite widely studied, at least in the United States, and some of the studies from Denmark have the same phenomenon.

Gurling

It is possible that it is nonspecific, but yet it is still very valuable in our family studies. I wonder if you could take us a bit further in terms of the mode of transmission. If you choose those sibships in which neither parent was affected and you presumably have incomplete penetrance in the parents, can you then identify one unaffected parent or the other as a sort of genetic index and determine whether the mode of transmission was dominant?

Blackwood

Because these families were not systematically ascertained, it is difficult to perform segregation analysis. But we are now looking at the mode of transmission.

Vogel

Do we have any evidence on the normal resting EEG? You know that in schizophrenics abnormalities have been described before and it would be interesting to know whether any deviations of the normal EEG pattern distinguish between your patients and controls and the unaffected relatives.

Blackwood

We have not analyzed the resting EEG of patients or their relatives.

Mendlewicz

Can you exclude the potential effect of treatment with psychotropic drugs or neuroleptics? And what about chronic treatment with neuroleptics?

Blackwood

We did a study recording P300 from 20 drug-free schizophrenics when they were acutely ill and again at follow-up at 1 month and at 2 years. There was no change in the P300 latency following treatment and recovery from acute symptoms at these follow-up periods. This supports other studies which show there are no long-term effects of neuroleptics on P300 latency or amplitude in schizophrenia. This does not rule out the possibility of some acute drug effects on patients or controls. In the family study, the symptom-free relatives were all drug-free.

Cloninger

I think the discussion we are having and the concern that has been expressed about the use of physiological or other markers that may be nonspecific in a population should be related back to the comment that Ming Tsuang made. That is, if we select our probands according to rather rigid categorical criteria, then we can be more flexible and can look at things within the family that in the general population do not have the specificity we need. But within families who have already been selected by something like one or two probands with the narrow phenotype we can use it. So the probability changes within these families and we can increase our sensitivity for relevant cases in a spectrum and maintain sufficient specificity by requirements for the index cases.

Cotransmission and Comorbidity of Affective Disorders, Anxiety Disorders, and Alcoholism in Families

W. Maier and K. R. Merikangas

The Problem

Family and twin studies have shown a lack of familial diagnostic homotypia. Many studies in probands with disorder A found not only A but also other disorders B to aggregate among the first-degree relatives as compared to control families. Some examples: The IOWA 500 Study (Tsuang et al. 1980) reported excess morbidity of bipolar affective disorder in families of schizophrenic probands. Gershon et al. (1988) found an aggregation of unipolar depression in families of schizophrenic probands. Farmer et al. (1988) observed major affective disorders in cotwins (monozygotic) of schizophrenic probands more often than expected by chance; more recent investigations such as Guze et al. (1985) and Slater and Cowie (1971) also found affective disorders to aggregate in families of schizophrenic probands. An excess morbidity of unipolar major depression in families of bipolar probands was often replicated (overview: Gershon et al. 1982); familial overlap between unipolar depression and alcoholism (Merikangas et al. 1985), between anxiety disorders and alcoholism (Noyes et al. 1986), and between unipolar depression and anxiety disorders (Munjack and Moss 1981) was also found. The reasons for this familial heterogeneity were not clarified by these family studies.

Virtually there are two main alternative explanations for the co-occurrence of different disorders in families: either different, distinct disorders A and B are cosegregating in the families or disorder A shares etiological factors with disorder B which are transmitted in families and therefore B might be considered as an alternative expression of A. The last-mentioned alternative is especially likely if (a) the excess morbidity of B is mainly encountered in relatives of probands who present with both disorders A and B (comorbidity in probands) but not in probands with A alone, (b) or in relatives who simultaneously report a lifetime diagnosis of A (comorbidity in relatives), or (c) in children of probands with A who are married to subjects with B (assortative mating).

The intention of this paper is to examine the familial relationship between bipolar and unipolar affective disorders, anxiety (panic disorders/agoraphobia) and alcoholism with regards to the alternative explanations. This kind of analysis is crucial to genetic research as far as phenotypes are concerned (e.g., in linkage studies): a multiplex family loaded with cases of disorder A might also show cases with disorder

B. It has to be decided whether B might be considered an alternative expression of A and such a decision might be guided by modes of analysis described in the following.

Models of Analysis

A simple criterion for the segregation of two etiologically distinct disorders A and B in families of probands with A is that the lifetime risk of A (but without B) and B (but without A) are elevated in probands with A but without B. If this criterion is fulfilled, neither comorbidity of A and B in probands nor comorbidity in relatives explains the simultaneous familial aggregation of A and B; assortative mating also stands to be ruled out as a source of diagnostic heterogeneity in families by comparing the lifetime risk of B in mates of probands with disorder A to the population rate for B.

More sophisticated approaches to decide between the two alternatives are offered by the application of statistical models.

Regression Models (Logistic Regression, Cox Regression Models)

Regression models identify factors (independent variables) which modulate the manifestation of a bipolar disorder in relatives. Occurrence of B in relatives (or: time to first manifestation of B or age at investigation if B always was absent) is considered the dependent variable and the proband, comorbidity for A and B, lifetime diagnosis of A in the relative as the main independent variables. In children of probands the diagnostic status of the mate of the proband could also be included in order to account for assortative mating; the weights allocated to the independent variables reflect their particular impact on the manifestation of disorder B in relatives (Lee 1980). Interaction between independent variables is controlled by changing the sequences of independent variables and comparing the extent to which they fit the model.

Cotransmission Analysis

If the familial transmission of A and of B is compatible with a multifactorial/ polygenetic mode of transmission, a bivariate multifactorial threshold model with one dimension for each disorder might be assumed. A dimension measures the liability for a particular disorder; it is subdivided into two components by a threshold, whereby subjects with vulnerability indices smaller than this threshold are not affected and the others are. The scores of liability are normally distributed in the general population and in the sample of relatives of patients. Given this model and given the observed prevalences of A, B, and A and B in the control sample and in families of probands with A, B, and A and B the proportion of variance transmitted in families (so-called heritabilities) and the pairwise correlation between transmissible and nontransmissible components of the two disorders may be estimated; these estimates define the "full model"; in order to identify the most parsimonious model particular coefficients can be fixed to zero and the other coefficients can again be estimated under this condition. Using the "nested models" technique the most parsimonious model (maximal number of coefficients put to zero) can be identified which fits the data not significantly ($p < .05$) worse than the "full model"; this model is called the "best fitting model." The procedure can easily be extended to three disorders A,B, and C reflected by three vulnerability dimensions. Whereas the logistic regression technique or the Cox regression model does not rely on a particular mode of familial transmission, the application of the last-mentioned cotransmission analysis requires that there be good reasons for assuming a multifactorial transmission for each of the disorders under study. Therefore the application of this model to bipolar disorder is questionable as a major locus effect is likely in this case (Rice et al. 1987).

With very few exceptions (e.g., Merikangas et al. 1991) the available literature on family studies of the disorders mentioned above holds with diagnostic hierarchies. The hierarchical approach proposes a gradation by severity between psychiatric disorders; the most severe lifetime condition is noted as are less severe conditions which might also be present. A distinction between "pure" and combined types of disorders A and B as required in identifying a causal link between A and B is not made here, but in most family studies the hierarchical approach is taken. Therefore these studies are not fully informative for the kind of analysis wanted for deciding between the aforementioned alternatives.

Analysis of Family Study Data

In order to compensate for methodological shortcomings and in an attempt to define the nature of the relationship of some of the disorders of interest, both types of analysis were applied to a recently completed family study conducted in Mainz, Federal Republic of Germany. Here, hospital patients ($n = 555$) were consecutively entered in the study who had been admitted with either schizophrenia, schizoaffective, or major affective disorders; 75 individuals with panic disorder or alcoholism but without lifetime diagnoses of any of the three aforementioned major disorders were included as probands; and 109 matched control probands were recruited from the general population. The control probands were not screened by psychiatric lifetime diagnoses as recommended by Tsuang et al. (1988). Probands and 81% of their first- degree relatives were directly interviewed by the semistructured clinical interview SADS-LA; family history information was collected for all relatives (dead or alive, available or not available). The interviews, family history information, and medical records were combined to receive a best-estimate lifetime diagnosis (Leckman et al. 1982). Direct interviews and diagnostic assessments of patients were

Table 1. Familial risk of schizophrenia and bipolar affective disorder by proband subtype

Morbid risk in relatives	Proband diagnosis			
	Schizophrenia only (%)	Schizophrenia and bipolar disorder (%)	Bipolar disorder only (%)	Controls (%)
Bipolar disorder	0.6	3.6	5.6	1.0
Schizophrenia + bipolar disorder	0.3	2.0	0.0	0.0
Schizophrenia	4.3	2.0	0.8	0.3

Underlining indicates elevated risk compared to control families (chi-square, $p = .05$).

performed blindly for the diagnostic status of relatives and vice versa in order to avoid any systematic bias in favor of diagnostic homogeneity. The prevalence rates displayed in the tables combine interviewed subjects and those with family history information only.

Familial Relationship of Mania and Depression

The majority of family studies found unipolar depression to aggregate in families of probands with unipolar depression and of probands with bipolar disorder, whereas bipolar disorder only aggregates in bipolar disorder probands (Rice et al. 1987). These replicated observations suggest that unipolar depression might be an alternative expression of bipolar disorder in families of bipolar probands. According to the proposed guidelines of analysis, familial aggregation of unipolar depression in probands with "pure" type mania (i.e., manic episodes without unipolar depression) is crucial for the discussion of the hypothesis that unipolar depression is an alternative expression of bipolar disorder; but the status of pure-type mania in the bipolar spectrum remains controversial (Angst et al. 1980). In more recent papers pure-type mania is subsumed under bipolar disorder. We obtained data in our family study indicating there is a morbidity risk of the pure type and the combined type of mania and depression in relatives of patients subclassified in the same manner and in relatives of controls.

Table 1 shows that relatives of probands with pure mania are at an increased risk for unipolar depression. This observation favors an etiological link between pure mania and pure depression if assortative mating can be excluded as the reason for this relationship. However, the mates of married probands with pure mania are at a slightly increased risk of unipolar depression (3 of 18, 17%) compared to the mates of controls (6 of 47, 12%); however, as sample sizes in both groups are limited this difference is not significant (Heun and Maier 1991). In particular with a limited sample size the morbid risk in children where one parent is suffering from pure mania and one from pure depression cannot be compared to that in children where only one of the two is suffering from pure mania.

These figures lend support to the hypothesis that unipolar depression in families of bipolar probands might be an alternative expression of bipolar disorder in families of probands with bipolar disorder. However, it cannot be fully excluded that the familial link between mania and unipolar depression is partly due to assortative mating. Assortative mating of this particular type was also found in the studies by Dunner et al. (1976) but not consistently so in other studies.

Bipolar Disorder and Alcoholism and Panic Disorder/Agoraphobia

In a recent review Merikangas and Gelernter (1990) concluded that it can be learned from family studies that bipolar and unipolar depression segregate independently in families although some, but not all, studies (e.g., Gershon et al. 1982) found comorbidity between both disorders in patients and therefore elevated familial risks for alcoholism in probands with bipolar disorder. The empirical basis for these observations is not strong, as the number of controlled family studies supporting this view is very limited — only two (Cloninger et al. 1979; Angst 1966). Therefore it might be useful to consider the figures emerging from our family study.

Relatives of probands with bipolar disorder are more often affected by alcoholism (12%) than relatives of controls (7.8%). The lifetime risks after breaking down probands and affected relatives according to "pure" and combined types of disorders are shown in Table 2.

Table 2 indicates very clearly that only those relatives are at an elevated risk of bipolar disorder (including bipolar I and bipolar II) who are related to an index proband with this disorder; the same is true for alcoholism. The combined proband group is at an elevated familial risk for both the pure and the combined type.

The Cox regression analysis (not shown in a table) also shows that the occurrence of alcoholism in relatives is modulated by male sex and alcoholism but not by bipolar disorder in probands (sample: relatives of the probands with bipolar disorder or alcoholism or relatives of the control probands; dependent variable: time to onset of alcoholism in relatives; independent variables: sex of relatives and probands,

Table 2. Familial risk of unipolar depression and bipolar disorder by proband subtype

Morbid risk in relatives	Proband diagnosis: Mania only (%)	Mania + depression (%)	Unipolar depression only (%)	Controls (%)
Unipolar depression	14.5	16.2	15.0	7.6
Mania + depression	4.0	6.0	0.9	1.0
Mania only	3.0	2.0	0.5	0.0

Underlining indicates elevated risk compared to control families (chi-square, $p = .05$)

diagnostic status of probands and relatives. Another particular feature was obtained by regression analysis: bipolar disorder in relatives also increased the likelihood of concurrent alcoholism. This means that once a relative is affectively ill he or she is likely to get addicted to alcohol. It was further shown that comorbidity in probands cannot explain this association by controlling this factor in the Cox analysis. The examination of morbidity risks in mates of probands reveals an increased rate of alcoholism in probands (10.5%) with bipolar disorder compared to mates of controls (7.6%; Heun and Maier 1991).

In conclusion there might be three reasons for the observed overlap of bipolar disorder and alcoholism in families: first, comorbidity of probands for both conditions might reflect the increased likelihood of seeking treatment if a proband is suffering from alcoholism and bipolar (especially bipolar II) disorder as compared to suffering from bipolar disorder alone; also assortative mating might contribute to the excess rate of alcoholism in families of probands with bipolar disorder. A third reason might be that relatives who are affected by bipolar disorder are at an increased risk of alcoholism independent of whether the proband is suffering from alcoholism. Common etiological factors transmitted in families can be ruled out as reasons for the coexistence of bipolar disorder and alcoholism.

As to the risk of anxiety disorders in relatives of bipolar probands reported in controlled family studies on bipolar disorders, there is divergent evidence for an aggregation of anxiety disorders in families of bipolar probands (e.g., Gershon et al. 1982; Merikangas and Gelernter 1990).

These ambiguous situations suggest that (a) the particular samples are different by the proportion of probands suffering from additional nonaffective disorders and that (b) the prevalence rates of anxiety disorders in relatives is moderated by presence or absence of comorbid conditions in probands. The results of our family study support this (for a more detailed analysis see Maier et al. 1991). Panic disorder and/or agoraphobia is aggregate in families, but only if the proband also presents with this disorder either in a pure or in a combined form. Therefore there is no indication of overlap between those etiological factors of bipolar disorder and panic disorder/agoraphobia which are transmitted in families.

Unipolar Depression, Alcoholism, and Panic Disorder

The familial pattern of the concurrence of these three disorders is complex, especially if diagnostic hierarchies are disregarded. The importance of the familial overlap of these three conditions is underestimated by most of the family study investigators as (a) in family studies including probands with unipolar depression the lifetime risk of alcoholism and panic disorder/agoraphobia in families is often not reported, and (b) there is not a single published report of a controlled family study in which patients with both unipolar depression and anxiety disorders (including pure types of both disorders) were simultaneously recruited. The most careful analysis of the patterns of overlap of unipolar depression, alcoholism, and anxiety disorders was recently performed in the Yale Family Study (Merikangas et al. 1991);

however, this family study is only based on probands with unipolar depression and controls. Whereas the familiality of unipolar depression has received a lot of attention, the nature of the familiality of alcoholism and, especially, anxiety disorders is less clear. But as regards the mode of segregation there is agreement that neither of the three aforementioned disorders fits a simple mendelian mode of transmission. Multifactorial/polygenetic models seem to be more appropriate, though their mathematical formulations (e.g., model of Reich and Falconer) are hard put to explain the sex-specific rates of disorders in relatives, given that the sex of the probands has no impact on the familial risks. Assuming the familial transmission of each of the three disorders fits the dimensional model of multifactorial transmission, the analysis of cotransmission as proposed by the group of N. Risch at Yale (Merikangas et al. 1988) might turn out to be very useful for the decision on quality and extent of communalities of corresponding transmissible familial factors. The most appropriate data set for this kind of analysis would be a controlled family study with probands suffering from either of these three conditions (pure forms as well as all combinations of them) and hierarchy-free diagnostic assessments of relatives. Our family study provides such material; the precise figures for pure and combined types of disorders are presented in detail elsewhere (Maier et al. 1992).

In summary, the prevalence rates in directly interviewed first-degree relatives — disregarding the distinction between pure and combined types — are:

The lifetime risk (percentage, not corrected for age) of

- Unipolar major depression to	
Relatives of probands with unipolar depression (without other diagnoses)	18.9%
Relatives of probands with other diagnoses (alcoholism, panic disorder)	12.9%
Relatives of control probands	7.6%
- Alcoholism to	
Relatives of probands with alcoholism (without other diagnoses)	17.0%
Relatives of probands with other diagnoses (unipolar depression, panic disorder)	12.0%
Relatives of control probands	7.8%
- Panic disorder or agoraphobia to	
Relatives of probands with panic disorder/agoraphobia (without other diagnoses)	5.0%
Relatives of probands with other diagnoses (unipolar depression, alcoholism)	3.1%
Relatives of control probands	2.2%

The results of this cotransmission analysis are demonstrated in Table 3: the best--fitting model (i.e., the most parsimonious model not differing from the full model

Table 3. Parameters of polygenic-multifactorial threshold model for cotransmission of depression, alcoholism, and anxiety

Model-dependent estimates of the components of variance

H_d	Transmissibility of depression
H_{al}	Transmissibility of alcoholism
H_{an}	Transmissibility of anxiety
$P_{d\text{-}al}$	Correlations between transmissibility for depression and alcoholism
$P_{d\text{-}an}$	Correlations between transmissibility for depression and anxiety
$P_{al\text{-}an}$	Correlations between transmissibility for alcoholism and anxiety
$-_{d\text{-}al}$	Correlations between environmental components for depression and alcoholism
$-_{d\text{-}an}$	Correlations between environmental components for depression and anxiety
$-_{al\text{-}an}$	Correlations between environmental components for alcoholism and anxiety

	Full model	Best-fitting model
H_D^2	0.630 ± .116	0.631 ± .114
H_A^2	0.522 ± .132	0.535 ± .128
H_P^2	0.493 ± .132	0.514 ± .129
$P_{D\text{-}A}$	0.0 —	0.056 ± .114
$P_{D\text{-}P}$	0.653 ± .160	0.689 ± .127
$P_{A\text{-}P}$	0.437 ± .174	0.616 ± .139
$-_{D\text{-}A}$	0.208 ± .178	(0)
$-_{D\text{-}P}$	0.096 ± .202	(0)
$-_{A\text{-}P}$	0.366 ± .216	(0)

by the degree of fitting the model) was received after the correlations between the nontransmissible components of the three disorders were ignored (i.e., correlations were fixed at zero); the full and the best-fitting model indicate strong components transmitted in families (H) which are all significantly different from O $(p<.005)$; these components may either be due to genetic or to environmental factors shared by all family members. The estimates of the correlations between the transmissible factors were high or moderate for unipolar depression and panic disorder and for panic disorder and alcoholism but not for unipolar depression and alcoholism.

It might be concluded that unipolar depression and alcoholism do not share etiological factors which are transmitted in families whether they are of genetic or environmental origin; the other two pairs of disorders have those factors in common and consequently depression and anxiety disorders may be expressions of the same etiological conditions, as may panic disorder and alcoholism.

Whereas the conclusion that unipolar depression and alcoholism have a clearly distinct familial background contradicts the "spectrum" concept of Winokur, the results are in agreement with the Yale Family Study (Weissman et al. 1982). In terms of statistical significance Merikangas et al. (1986, 1991) reported the identical pattern of cotransmission between depression, anxiety, and alcoholism (strong familial components of all three disorders and significant correlations between the transmissible factors of unipolar depression and anxiety disorders and of anxiety disorders and alcoholism but not between unipolar depression and alcoholism). This

Table 4. Familial risk of bipolar disorder and alcoholism by proband subtype

Morbid risk in relatives	Proband diagnosis: Bipolar disorder only (%)	Bipolar disorder and alcoholism (%)	Alcoholism only (%)	Controls (%)
Alcoholism	9.0	12.5	17.0	7.8
Bipolar disorder + alcoholism	3.5	3.6	0.0	0.0
Bipolar disorder	4.9	6.0	0.5	1.0

Underlining indicates elevated risk compared to control families (chi-square, $p = .05$)

similarity is even more surprising as both family studies differed in three major aspects: (1) in the Yale family study, besides the controls, only probands with unipolar depression but not with "pure" anxiety disorders or "pure" alcoholism were investigated; (2) the present analysis of the Mainz family study includes only panic disorder/agoraphobia but not generalized anxiety disorders or other phobias; and (3) the family structures (number of first-degree relatives per family is about twofold higher in the Yale family study) and the percentage of interviewed living relatives (80% in the Mainz study and substantially less in the Yale study). Given these divergent patterns of the study designs the similarity of the results of both studies lends strong support to the validity of their conclusions (Table 4).

These results also indicate that the comorbidity between unipolar depression and panic disorder/agoraphobia and between panic disorder/agoraphobia and alcoholism which are often encountered in clinical settings are not due to artifacts (e.g., treatment seeking behavior increases if the number of concurrent disorders in a subject increase).

Implications for Linkage Analysis

Linkage analysis examines the cosegregation of a phenotype with a genetic marker. The definition of the range of the phenotype conventionally indicates whether a particular psychopathological pattern is regarded as affected or not (categorical approach; dimensional approaches have also been proposed, but they are not taken into consideration here). Four minimal requirements for considering disorder B an alternative expression of the familial basis of disorder A in families of probands with A seem to be reasonable :

1. B aggregates in families of probands with A.

2. It can be excluded that the concurrence of A and B in probands explains the aggregation of B in families of probands with A (i.e., "pure-type" B should aggregate in families of probands with "pure-type" A); this possible explanation can be excluded by assessing lifetime risks of "pure-type" B in relatives of "pure-type" A probands.

3. The aggregation of B in relatives is not due to the presence of A in relatives (i.e., comorbidity of A and B in relatives). This possible explanation is ruled out by regression models with manifestation of B in relatives as the dependent variable and diagnostic status (affected with A) in relatives as the independent variable (no significant contribution of this independent variable).

4. This aggregation of B in families of probands with A is not due to assortative mating between subjects with A and subjects with B.

In families of probands with bipolar affective disorders which are a major locus of linkage studies we were able to show that unipolar depression might be considered an alternative expression of the bipolar disorder. However, neither alcoholism nor panic disorder/agoraphobia can be considered an alternative expression of bipolar disorder.

Conclusion

The analysis of cotransmission and comobordity as discussed in this paper can contribute to two major issues of the concurrence of two syndromes within subjects and families:

1. To decide whether two syndromes occurring within families more frequently than expected by chance have etiological factors in common or whether they indicate two disorders of different etiology.
2. Two decide whether comorbidity of syndrome as often observed in clinical settings is due to a common familial etiology or whether comobordity is likely to be due to artificial factors (e.g., has been induced by different treatment seeking behavior).

The analyses of our family study data suggest common familial factors for mania and unipolar depression and panic disorder/agoraphobia as well as for panic disorder/agoraphobia and alcoholism. Other pairwise combinations of the syndromes under study might also occur more frequently than expected by chance either in probands or in families; however, these putative excess morbidity rates are not due to familial factors shared by both components.

References

Angst J (1966) Zur Aetiologie und Nosologie endogener depressiver Psychosen. Monogr Gesamtgeb Neurol Psychiatr (Berlin) 112:1-118

Angst J, Frey R, Lohmeyer B, Zerbin-Rüdin E (1980) Bipolar manic-depressive psychoses: results of a genetic investigation. Hum Genet 55:237-254

Cloninger CR, Reich T, Wetzel R (1979) Alcoholism and affective disorders: familial association and genetic models. In: Goodwin DW, Erickson CK (eds) Alcoholism and affective disorders: clinical, genetic, and biochemical studies. SP Medical and Scientific Books, New York, pp 57-86

Falconer DS (1965) The inheritance of liability to certain diseases, estimated from the incidence among relatives. Ann Hum Genet 29:51-76

Farmer AE, McGuffin P, Gottesmann II (1987) Twin concordance for DSM-III schizoprehnia: scrutinising the validity of the definition. Arch Gen Psychiatry 44:634-641

Gershon ES, Hamovit J, Guroff JJ, Dibble E, Leckman JF, Sceery W, Targum SD, Nurnberger JI, Goldin LR, Bunney WE (1982) A family study of schizoaffective, bipolar I, bipolar II, unipolar, and normal control probands. Arch Gen Psychiatry 39:1157-1167

Guze SB, Cloninger R, Martin RL, Clayton P (1983) A follow-up and family study of schizophrenia. Arch Gen Psychiatry 40:1273-1276

Lee E (1980) Statistical methods for survival data analysis. Wadsworth, Belmont

Maier W, Lichtermann D, Minges J, Franke P, Heun R (1991 A controlled family study in panic disorder. J Psychiatr Res (in press)

Maier W, Lichtermann D, Hallmayer J (1992) The cotransmission of depression, alcoholism and panic disorder in families. (submitted)

Merikangas KR, Gelernter CS (1990) Co-morbidity for alcoholism and depression. Psych Clin N Am 13: 613-632

Merikangas KR, Leckman JF, Prusoff BA, Pauls DL, Weissman MM (1985) Familial transmission of depression and alcoholism. Arch Gen Psychiatry 42:367-372

Merikangas KR, Risch NJ, Merikangas JR, Weissman MM, Kidd KK (1988) Migraine and depression: association and familial transmission. J Psychiatr Res 22:119-129

Merikangas KR, Risch NJ, Weissman MM (1991) Comorbidity and co-transmission of alcoholism, anxiety and depression. Arch Gen Psychiatry (in press)

Munjack DJ, Moss HB (1981) Affective disorder and alcoholism in families of agoraphobics. Arch Gen Psychiatry 38:869-871

Noyes R Jr, Clancy J, Crowe RR, Hoenk PR, Slymen DJ (1978) The family prevalence of anxiety neurosis. Arch Gen Psychiatry 35:1057-1059

Reich T, Rice J, Cloninger CR, Wette R, James J (1979) The use of multiple threshold and segregation analysis in analyzing the phenotypic heterogeneity of multifactorial traits. Ann Hum Genet 42:371-389

Slater E, Cowie V (1971) The genetics of mental disorders. Oxford University Press, Oxford

Tsuang MT, Winokur G, Crowe R (1980) Morbidity risks of schizophrenia and affective disorders among first degree relatives of patients with schizophrenia, mania, depression and surgical conditions. Br J Psychiatry 137:497-504

Tsuang MT, Fleming JA, Kendler KS, Gruenberg AS (1988) Selection of controls for family studies. Arch Gen Psychiatry 45:1006-1008

Weissman MM, Kidd KK, Prusoff BA (1982) Variability in rates of affective disorders in relatives of depressed and normal probands. Arch Gen Psychiatry 39:1397-1403

Discussion of Presentation W. Maier and K. R. Merikangas

Tsuang
Would you define what you mean by comorbidity.

Maier
A and B are comorbid when they both occur either within the same episode simultaneously or in different episodes. Now, of course the question is relevant whether the simultaneous or separate occurrence of both symptoms has any influence on this kind of analysis. When looking more precisely at the data, indeed, the differences are limited. At least the conclusions remain the same. That means you can consider schizomania and patients with schizophrenia who have also had a manic episode more or less as the same cases with regards to their families.

Tsuang
Do you have enough probands to analyze those in which A took place first and the B subsequently and vice versa.

Maier
I have enough cases for considering the impact of simultaneous and nonsimultaneous occurrence of two syndromes. But I do not have enough cases for, in addition, addressing the question of whether the primacy of A or B is of any impact. The cases with mania occurring before schizophrenia are rare, about 10% or 20% of the cases of schizophrenia. And so we have here about 20 cases, which is not enough.

Mendlewicz
The prevalence of schizophrenia is about 1% and the prevalence of bipolar disorder is about 1.5%. In your experience what is the frequency of the comorbidity between bipolar disorder and schizophrenia? It should be very rare.

Maier
In the general populations it is negligible. The proband data I presented are from patients, and 20% of probands with schizophrenia present with a manic syndrome at least once during their lifetime.

Mendlewicz
You have a proband in the comorbidity group who has both bipolar disorder and schizophrenia. That is very rare according to our experience.

Maier
Yes, it is about 30 cases within the schizophrenic sample of 150 schizophrenics and in addition we investigated 50 cases of schizomania. On the other hand there is not a single case in the sample of control families with simultaneous schizophrenia and mania.

Ackenheil
If you consider schizophrenia and mania to be different entities and if the prevalence in the general population is only 1%, then comorbidity of two different nosological entities in this sample would be a very seldom occurrence.

Maier
It does not occur in the control population but in those probands who are patients. Huber, for example, describes for the German population of schizophrenic patients, 20% manifesting first with cyclothymia and then with schizophrenia in different episodes. A and B are syndromes: they may occur simultaneously or in different episodes. This temporal pattern of occurrence, simultaneous or separate, has no major impact on the results of this analysis.

Hippius
On the one hand, looking at your figures, one has to think as Dr. Mendlewicz that the concept of comorbidity is comorbidity of two different diseases. On the other hand, what is your concept behind this. Are you returning to the old concept of the "Einheitspsychose"? I think these are the two positions and the two concepts should be allowed. For planning genetic research in the future, we should know whether we should follow the new edition of "Einheitspsychose" or the concept of real comorbidity. I would like to have your comments on this.

Maier
First, I have to make a comment on the term comorbidity. The term comorbidity was introduced by the group of Myrna Weissman into the field of psychiatry. Strictly speaking the word is not quite correct. The term is being used to mean co-occurrence of syndromes either during the same episode or at totally different time points. Additional analysis shows that the distinction between coexistence of the two syndromes during an episode and co-occurrence within lifetime has no impact on the familial risk figures.

Hippius
I have some sympathy with your findings, because I remember working with Jules Angst where each of us investigated 100 patients in the Berlin and Zürich hospitals

who had been hospitalized more than 10 times. When we compared the diagnoses of these 200 cases, the changes from one diagnosis to another were so disturbing that we stopped our investigation, although it was an early investigation of comorbidity.

Maier

I would like to stress that the transition probability from schizophrenia to pure-type mania and back is rather low. This is a transition of lowest probability. The transition from schizophrenia to unipolar depression has a high probability.

Ackenheil

Do you mean the transition of syndromes now, from schizophrenic to manic syndromes, because there is now again some confusion.

Hippius

"Hypomane Endzustände Buerger-Prinz": is that what you have in mind?

Maier

Yes, this together with the description of Huber, mania before schizophrenia onset.

Cloninger

Can you tell us, in your data is there an excess of schizophrenia in the relatives of nonpsychotic unipolar depressives?

Maier

No, there is not an increased rate for schizophrenia in the families of unipolar depressive individuals. Whether they are psychotic depends on how broad your concept of psychotic disorder is.

Cloninger

I think it is misleading to say that you should include unipolar depression in the schizophrenia spectrum if in fact nonpsychotic major depressives do not have an excess of schizophrenic relatives. If you would make that one revision to your final conclusion I would feel very comfortable.

Maier

For a disorder to be included in the schizophrenic spectrum I would not require that the family members of this particular condition show excess morbiditiy for schizophrenia. Take, for example, schizotypal personality, which is considered a constituent of the schizophrenia spectrum. The findings with regard to the excess morbidity of schizophrenia in probands with schizotypal personality disorder are extremely controversial and are not reflected by the readiness with which clinicians include schizotypal personality disorder in the schizophrenic spectrum. And it is totally compatible, when thinking along the lines of the Reich-Falconer models of multifactorial transmission, that a mild disorder on the dimension of liability does not show an excess morbidity for the most severe condition on that dimension.

Cloninger

I would require that it be made part of the spectrum, because it says that the cases in the general population are so nonspecific that only a minority of them are related. I have actually made that criticism for schizotypal personality being part of the schizophrenics spectrum, too.

Maier

Then we have different criteria for defining a schizophrenic spectrum. The choice of how to define a spectrum is more a matter of convention than one of science.

Tsuang

It is probably very dangerous to base comorbidity diagnoses from the cross-sectional point of view. Probably follow-up needs to take place to see the stability of this core syndrome and which may not be the final stable diagnosis. As a clinician you know some schizophrenics may develop a manic phase, a depressive phase, and then gradually turn out to be schizophrenic. And some with schizophrenia later turn to alcohol. So for those types we cannot take comorbidity with a cross-sectional diagnosis of probands to interpret the results. Of course it is very meaningful from the preliminary findings, but to make this more meaningful long-term follow-up is important.

Maier

Yes, we are doing follow-up studies in some of the patients and some of the relatives. Analyzing the subsample separately of those relatives having a 2-year follow-up, reassessment twice a year provides no reason to doubt the validity of the conclusion. Of course I cannot exclude in principle your criticism.

Mendlewicz

Just one point on the issue of unipolar depression. A number of studies were done in the 1950s on melancholia. One of those is the study of Kallmann, which indeed shows an increased rate of schizophrenia in relatives of melancholic patients. When you look at the case descriptions, these patients did not have psychotic depression, but fit more the old melancholia definition of the French and German nosology. And there is a French study, too, reporting a higher rate of schizophrenia in the relatives of melancholic subtypes.

Crow
I think the discussion shows the sort of troubles we are in nosologically. But as a practical issue, what do you recommend for linkage studies. Ming Tsuang is going to draw a single line down the middle of psychosis, and schizomanics are going to go one way and schizodepressives are going to go another way. The question is, how many lines are you going to draw, how comfortably can you draw them, and where are they going to be drawn?

Maier
I draw two lines, coming up with three classes: pure bipolar; schizomania, whereby schizomania means either within the same episode or mania and schizophrenia in different episodes; and the pure schizophrenic group. But these classes mean that our evidence is not so great that we can combine those three classes. It may turn out that you can combine them but currently caution is warranted.

Kidd
Related to Dr. Mendlewicz comments at the very beginning I would like to point out how common schizoaffective types, the comorbid types, seem to be in your patient sample. Because it is very well recognized in medical epidemiology that patients tend to have more than one disease, those with more than one disease are the ones most likely to seek attention and most likely to get into a clinical situation, so they are probably disproportionally represented among the clinical patients compared to what they are in the real population. I do not see anything surprising about what you found.

Vogel
On the one hand you do not find an increase of schizophrenia in relatives of unipolar patients. On the other hand you do find an increase of unipolar patients among relatives of schizophrenics. This is of course a contradiction which has to be explained. In my opinion the explanation is simply the following. Whereas schizophrenia is relatively rare in the population, unipolar psychoses are much more common. According to your data, almost 10% of our population has had at least one phase that could be regarded as clinically important unipolar depression. When you look at this, then there are vastly more probands that have just unipolar disease than unipolar patients who belong to the families with schizophrenia. I think, this simple fact, this difference in prevalence between the two conditions by a factor of ten might explain it.

Maier

To comment on this, you could also discuss this question by extending the boundaries of schizophrenia. Given the validity of the schizophrenic spectrum, you could consider schizotypal personality disorder and perhaps schizoid personality disorder and unspecific psychoses to be variants of schizophrenia. When you do that, then you can increase the base rate substantially, and with regard to the base rate in the neighborhood of unipolar depression. But when doing that, you will not find an increased rate for this extended concept of schizophrenia or psychoses in the families of probands with nonpsychotic unipolar depression, not in those showing the subtype of melancholia either. My explanation is based on a multifactorial multiple threshold model; if you have a rather severe illness on the dimension of vulnerability, it is compatible with a model that mild conditions do not show excess familial morbidity with regards to the most severe condition on that continuum.

Reich

I would like to hear some comments from our senior German psychiatrists on the utility of the various traditional diagnoses, which I believe in some of the earlier nosological systems included things like affect latent paraphrenia and nonsystematic schizophrenias. I understand those have gone out of fashion but perhaps might come back into fashion.

Hippius

That is a German speciality by Leonhardt, but it is customary to smile a little bit about Leonhardt. I think that the way he went about identifying more nosological entities, ultimately there were more than 15, was wrong. But on the other hand, we could learn much from his definition of syndromes. We speak only of schizophrenia and affective psychoses. It would be a big advantage if we included in genetic research Leonhardt's concept — not as entities but as a preliminary step in the operationalization of differentiating the symptomatology and the syndromes. Therefore, I think the concept schizoaffective psychosis is the first step in the Leonhardt direction. Leonhardt was the second psychiatrist after Kraepelin to look into the course of the illness independent from the symptomatology. He already differentiated between unipolar and bipolar disorder. He was very autistic with his research, but it would be worthwhile to revive the psychopathology.

Vogel

I wish to remind you that the problem we are discussing now is a very old one in psychiatric nosology. One of the classics of psychopathology is a book of Jaspers *General Psychopathology*. In this book Jaspers explains and sets out the problem which we are discussing here very clearly. And he even suggests that patients should not be classified according to diagnosis such as schizophrenia or schizoaffective, but that within a psychiatric unit the patient should be compared with other patients and then an empirical classification like this be made. Of course I do not advocate this procedure, but it shows you, that this problem of nosology or the limits of different diagnoses is a very old one and nobody has come to any conclusion. For me personally it is a reason never to plan a linkage study with psychiatric diseases.

Hippius
To read and study Jaspers with the aim of developing an operationalization of psychiatric phenomena on the basis of Jaspers is very difficult. As Luxemburger has already stated, "Schizophrenia is a good working hypothesis but not more." Jaspers' writings are full of these statements but he was not very successful in the operationalization of psychiatric diagnoses and symptomatology. Jaspers was not only a psychiatrist, he was a philosopher, too; that is the difference.

Maier
Finally, I want to stress that it is not sufficient in linkage studies to stick to the clinical phenomenology. This could in some way be extended by neuropsychological, neurophysiological or biochemical methods.

Biological Marker Studies in Families*

B. Bondy, M. Ackenheil, M. Ertl, H. Giedke, P. König, C. Mundt, H. Sauer, and G. Schleuning

Controversy has arisen from conflicting results obtained with new molecular techniques to unravel the genetics and biology of major psychoses. Etiologic heterogeneity can certainly explain part of the discrepancy. The main reason might be due to difficulties in defining the clinical phenotype. Furthermore, phenocopies or unexpressed genotypes complicate the problem of diagnoses. Several investigators have proposed that the identification of phenotypical vulnerability traits might be a promising way out of this dilemma (Tsuang et al. 1987; Erlenmeyer-Kimling and Cornblatt 1987; Belmaker 1991). The power of molecular genetics, applied in families homogenous with regard to a biological parameter, could thus be intensified. However, despite intensive research, the identification of vulnerability traits has been rare. Besides psychophysiological parameters, which are extensively discussed by Blackwood and Muir (this volume), biochemical investigations may also have some benefit as phenotypical markers for major psychoses.

We report here the results of studies on spiperone binding capacity of mononuclear cells in families of schizophrenic and schizoaffective index probands. Binding of the dopamine antagonist spiperone to mononuclear cells was found to be increased in a number of schizophrenic individuals (LeFur et al. 1983; Bondy et al. 1984; Rotstein et al. 1983; Grodzicki et al. 1990), stable over time (Bondy and Ackenheil 1987), and genetically determined (Bondy et al. 1990a). These results hold promise in the identification of families homogenous for a heritable biological parameter.

Methods

Subjects

A total of 22 index probands (9 women, 13 men, aged 22-55 years) were recruited from 5 different clinical centers (Psychiatric Hospitals of the Universities of Munich, Tübingen, and Heidelberg, FRG, and the local district hospitals in Haar/Munich, FRG, and Valduna, Austria). Included were index probands with the diagnosis of

*This work was supported by grants from the Deutsche Forschungsgemeinschaft.

schizophrenia (n = 14) and schizoaffective disorder (n = 5). The results were compared to those of bipolar (n = 2) and unipolar disorder (n = 1). Preconditions for inclusion were at least one additionally affected relative and participation of minimum of four further relatives. Twenty index probands were under medication at the time of investigation. Of 336 recruited relatives 166 were personally interviewed and agreed to venipuncture for the investigation of lymphocyte spiperone binding capacity. In 170 relatives, who refused to participate, diagnosis was made using the RDC-FH form.

Diagnoses

Diagnoses were achieved according to RDC and DSM-III criteria using the semi-structured interview SADS-LA (Spitzer und Endicott 1975), part of the CIDI (Robins et al. 1989) and SKID (Wittchen et al. 1987) to further characterize schizophrenia and personality. Patients with psychotic symptoms lasting longer than 2 months, residual symptoms, or those on chronic neuroleptic medication were classified as chronic; those with psychotic episodes in the range between 2 weeks and 6 months without symptoms of the defect were classified as acute.

Laboratory Investigation

Using 10 m*M* ethylenediaminetetra-acetic acid (EDTA) as anticoagulant 50 ml blood was drawn from each proband by venipuncture. Mononuclear cells were separated within 24 h after venipuncture and binding assays were performed with intact cells and the dopamine antagonist ^{3}H-spiperone and 1 μ*M* [+]-butaclamol as displacing agent as previously described (Bondy et al. 1990b).

Results

Clinical and biochemical parameters were combined in families of schizophrenic (n = 14) and schizoaffective (n = 5) index probands. Additionally included were families of bipolar (n = 2) and unipolar (n = 1) index probands. Of the 336 relatives 166 were personally interviewed and participated in the biochemical investigation. Most families included between 5 and 15 relatives; 4 large pedigrees (20-50 relatives) were additionally investigated. In total there were 221 relatives of schizophrenic index probands, 73 of schizoaffective, 13 of bipolar, and 29 of unipolar probands.

Clinical Results

The general incidence of psychiatric disturbances in relatives was rather high (n = 96; 28.5%), but did not differ between schizophrenic (28.9%) and schizoaffective

(27.3%) index probands. Quite interestingly, homotypia in the diagnosis schizophrenia was only observed in two smaller families, each comprising five probands. In the majority of families the variability of clinical diagnoses seemed to increase in parallel with the family size. In Table 1 the different diagnoses are listed for all relatives. Although schizophrenia and schizoaffective psychoses had a rate of 5.4% in the families of the respective index probands, unipolar major depression was the disorder most frequently observed among relatives of schizophrenic (11.3%) and schizoaffective (6.8%) probands.

Spiperone Binding Capacity in Mononuclear Cells

Based on earlier findings (Bondy et al. 1984, 1987) a B_{max} value above 4 fmol/10^6 cells was taken as the cutoff point for normal or increased capacity. The results revealed that in 17 (12 schizophrenic, 4 schizoaffective, 1 bipolar) of the 22 families, spiperone binding capacity was increased in index probands as well as in all psychiatrically affected relatives, independent of the clinical diagnoses. In these families also a proportion of phenotypically unaffected probands (14%) had increased binding capacity similar to that of the proband. The increased B_{max} values were in the range from 5 to 14 fmol/10^6 cells, thus being significantly above normal control values (0.8 to 3.7 fmol/10^6 cells). In five families we did not observe increased spiperone binding capacity either in the index probands or in any of the relatives. In three of these families the index proband was schizophrenic or schizoaffective. The remaining two families consisted of one pure bipolar and one large pedigree heavily loaded with unipolar major and minor depression. Figure 1 demonstrates a selection of pedigrees together with the spiperone binding capacity.

Table 1. Diagnoses in relatives

Diagnoses in relatives	Total (n=22)	Index probands SCH (n=14)	SA (n=5)	BP (n=2)	UP (n=1)
Schizophrenia	14	12	2	0	1
SA, mainly schizophrenic	5	3	2	0	0
SA, mainly affective	4	2	2	0	1
Bipolar disorder	16	3	0	0	4
Unipolar disorder	36	25	5	4	0
Personality disorders	9	5	1	3	0
Addiction	5	4	1	0	0
Suicide	7	3	4	0	0

Included were 336 relatives of 22 indexprobands. Psychiatric disorders were monitored in 96 probands. SA, schizoaffective disorder; SCH, schizophrenia; BP, bipolar disorder; UP, unipolar disorder

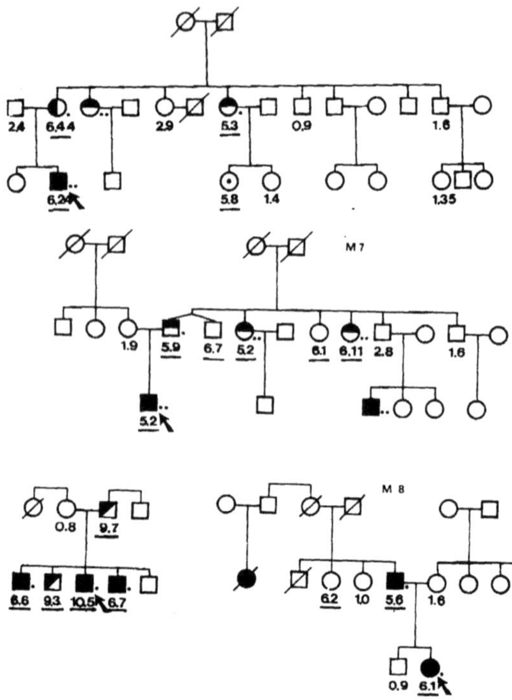

Fig. 1. Four pedigrees and spiperone binding capacity studied for each individual. Values above 4 (*underlines*) are classified as increased.□,○ male, female;■schizophrenic;◪ bipolar disorder; ⬒ unipolar disorder; ◧schizoaffective psychosis:⊡ Briquets syndrome

Discussion

In an earlier study we demonstrated that increased binding capacity of mononuclear cells for spiperone goes together with schizophrenia in families (Bondy and Ackenheil 1987). Since it turned out in recent years that multiple cases of psychiatric disturbances within one family might be even more frequent than homotypia, we investigated this potential vulnerability trait for schizophrenia in larger pedigrees with high loading for psychiatric disturbances. The main interest was focused on families with schizophrenia or schizoaffective psychoses. For comparison bipolar and unipolar cases were included as well.

Diagnostic homotypia was a rare phenomenon in our study. Although the diagnosis of the index probands predominated among affected relatives, other disturbances were quite common. Unipolar depression was found in excess, independent of the index case. But also major affective disorders appeared to be frequent in families of schizophrenics. These results correspond to recent findings (Gershon et al. 1988; Kendler et al. 1986) and again suggest how difficult it might be to separate the two conditions — schizophrenia and affictive disorders — completely on a genetic basis.

Concerning the increased spiperone binding capacity in mononuclear cells, initially proposed as a vulnerability marker for schizophrenia (Bondy and Ackenheil 1987), we have now observed that this phenomenon separates families into positive or negative for this finding. Increased binding capacity was not only found in families homotypical for the diagnosis of schizophrenia but also in most families with a variety of clinical diagnoses. In these families also 14% of phenotypically nonaffected relatives had increased binding capacity. In the families of two schizophrenic and one schizoaffective index patients, normal binding has been observed in all probands. From a clinical point of view, these families did not differ from those with positive results.

Of course these results raise a number of questions. One attractive model to explain the data is certainly the concept of continuum psychosis (Crow 1986). Another explanation might be that major psychoses represent different manifestations of one single gene, as proposed by Ridley and Baker (1990). These authors concluded on the basis of multiple clinical observations of Huntington's chorea that the phenotypic effect of a single gene might cover a wide continuum and major changes in phenotypic expression from one generation to the next. Besides these considerations, we would like to suggest that our findings likely increase the power of linkage analyses, since, using this paradigm, families can be identified that are homogenous for a biological trait beyond the nosological entity.

References

Belmaker RH (1991) One gene per psychosis? Biol Psychiatry 29:415-41

Bondy B, Ackenheil M (1987) 3H-spiperone binding sites in lymphocytes as possible vulnerability marker in schizophrenia. J Psychiatr Res 21(4):521-529

Bondy B, Ackenheil M, Birzle W, Elbers R, Fröhler M (1984) Catecholamines and their receptors in blood: evidence for alterations in schizophrenia. Biol Psychiatry 19(10):1377-1393

Bondy B, Ackenheil M, Ertl M, Peuker B (1990a) Genetische Untersuchungen zur Spiperon-Bindung an Lymphozyten. In: Saletu B (ed) Biologische Psychiatrie. Thieme, Stuttgart, pp 498-500

Bondy B, Ackenheil M, Engel RR (1990b) Methodology of 3H-spiperone binding to lymphocytes. J Psychiatr Res 24(1):83-92

Crow TJ (1986) The continuum of psychoses and its implication for the structure of the gene. Br J Psychiatry 149:419-429

Erlenmeyer-Kimling L, Cornblatt B (1987) High risk research in schizophrenia: a summary of what has been learned. J Psychiatr Res 21(4):401-41210

Gershon ES, DeLisi EL, Hamocit J, Nurnberger JI, Maxwell ME, Schreiber J, Dauphinais D, Dingmann CW, Guroff JJ (1988) A controlled family study of chronic psychoses. Arch Gen Psychiatry 45:328-336

Grodzicki J, Pardo M, Schved G, Fuchs S, Kanety H (1990) Differences in 3H-spiperone binding to peripheral blood lymphocytes from neuroleptic responsive and nonresponsive schizophrenic patients. Biol Psychiatry 27:1327-1330

Kendler KS, Gruenberg AM, Tsuang MT (1986) A DSM III family study of nonschizophrenic psychotic disorders. Am J Psychiatry 143:1089-1105

LeFur G, Zarifian E, Phan T, Cuche H, Flamier A, Bouchami F, Burgevin MC, Loo H, Gerard A, Uzan A (1983) 3H-spiroperidol binding in lymphocytes: changes in two different groups of schizophrenic patients and effect of neuroleptic treatment. Life Sci 32:249-255

Ridley RM, Baker HF (1990) Variable expression in the functional psychoses. A comparison with Huntington's disease. Schizophr Res 3:201-210

Robins LN, Wing J, Wittchen HU, Hetzer JE, Babor TF, Burke J, Farmer A, Jablenski A, Pickens R, Regier DA, Sartorius N, Towle LH (1989) The composite international diagnostic interview: an epidemiologic instrument suitable fur use in conjunction with different diagnostic systems and in different cultures. Arch Gen Psychiatry 45:1069-1077

Rotstein N, Mishra RK, Singal DP, Barone D (1983) Lymphocyte ^{3}H-spiroperidol binding in schizophrenia: preliminary findings. Prog Neuropsychopharmacol Biol Psychiatry 7:729-732

Spitzer RL, Endicott J (1975) Schedule for affective disorder and schizophrenia. Biometrics Research, New York State Psychiatric Institute, New York

Tsuang MT, Lyons MJ, Faraone SV (1987) Problems of diagnoses in family studies. J Psychiatr Res 21(4): 391-399

Wittchen HU, Zaudig M, Schramm E, Spengler P, Mombour W, Klug J, Horn R (1987) Strukturiertes klinisches Interview für DSMIII. Beltz, Weinheim

Discussion of Presentation B. Bondy et al.

Cloninger
You mentioned that your probands often were drug naive. But I assume that affected relatives sometimes have used drugs. So could you tell us just what is the effect of prior drug use on subsequent studies on the binding parameters.

Bondy
The first results I showed you were from the initial investigation with untreated schizophrenics; most of them were drug naive. In the family study I presented, most of the schizophrenics and a large proportion of the relatives have been under neuroleptic medication for between 2 months and 6 years. According to our previous studies, there is no influence of neuroleptic medication on the spiperone binding capacity, since the values remained on a constant level from acute psychoses, during neuroleptic treatment and in drug-free remission periods. It seems to be a stable phenomenon.

Maitre
If neuroleptic drug treatment does not influence spiperone binding, can you maybe suggest what kind of proteins are involved in the spiperone binding. It does not resemble any of the brain receptors which are labeled by spiperone.

Bondy
That is true. The initial concept of that investigation was that a dopamine receptor might be present on lymphocytes as a model for brain dopamine receptors. In the meantime it turned out that this is not the case, especially for the dopamine receptor types we knew until recently. According to our recent investigations we suggest the spiperone either labels or is specifically taken up by a specific dopamine transport protein in lymphocytes. We could show that lymphocytes have a specific transport system, and the dopamine uptake into lymphocytes can be inhibited by spiperone very effectively. And for about half a year we have been investigating spiperone binding and dopamine uptake in parallel in patients and controls and there is a very good correlation between these two parameters. Dopamine uptake is also a very stable phenomenon.

Gurling
Have you looked at the effect of smoking on spiperone binding?

Bondy

We have not figured that out precisely, but we do not think there is an effect of smoking on spiperone binding. We now have investigated more than 150 normal controls with a wide interindividual variance. The intraindividual variance is, however, rather small with an ICC of 0.89.

Kidd

Does the binding to lymphocytes involve T cells or B cells, and if B cells, have you looked at transformed lymphoblastoid cell lines?

Bondy

The specific binding to lymphocytes is mainly to B cells, about 90%; binding to T cells makes up about 10%. In patients we did not find any difference in the absolute number of B cells as compared to controls, thus this cannot be the reason for the increase. We have tried to do this assay in EBV-transformed cells, but unfortunately we have not succeeded in this; binding is rather low. And we suggest since it is not binding to a receptor but taken up, the membrane alteration that occurs in transformed cells might be the reason.

Tsuang

Just as a matter of interest, is there any hint of any differences in terms of clinical manifestations or family history in those binding positive versus those negative?

Bondy

There is certainly a hint of differences between positive and negative families, but we do not have enough negative families now. The schizophrenics in the negative families do not improve with neuroleptic medication. So far we have not precisely been able to figure out any difference in age at onset or sex. So we still have to be very careful in making final statements and have to extend the time of observation. In one of these negative families, there was a schizophrenic woman who had a bipolar child. This was the only family where it was the other way around. In all other families with mixed diagnoses we always found schizophrenics among the offspring of bipolar individuals.

Mendlewicz

Your binding observations show to some extent a genetic relationship — the twin data, the family data. On the other hand, you show that there is also a relationship to psychotic illness, because values are higher in schizophrenics than in controls. It seems to cluster in families. Have you done a quantitative genetic analysis to explore the relative importance of this?

Bondy
We have done not an analysis yet. This will be the next urgent step.

Twin Research

A. Bertelsen

Introduction

Twin research has over the last 70 years played an important role in psychiatric genetics, mainly by demonstrating the existence of a genetic factor through observations of higher concordance for monozygotic than for dizygotic twins. In the recent era of molecular genetics, twin research may appear outdated, as an almost historical approach. Twin research, however, is still contributing important information to research on the molecular level, too.

Twin research can contribute in the following five areas of major importance for molecular genetics:

1. Demonstration of a genetic factor
2. Identification of subgroups with high heritability
3. Evaluation of penetrance and expressivity
4. Assessment of phenotypic heterogeneity
5. Estimation of etiological heterogeneity

In psychiatric genetics twin research contributions in these areas will be exemplified with results from studies on the major psychiatric disorders: schizophrenia, manic-depressive disorders, and the so-called schizoaffective psychoses.

Schizophrenia

A considerable number of studies on twin series have without exception demonstrated the existence of a strong genetic factor in schizophrenia. The studies can be divided into a group of older studies and a group of more recent studies (Gottesman and Shields 1982). The concordance for schizophrenia and schizophrenialike psychosis showed higher monzygotic rates in the older series than in the newer ones, ranging from .58-.69 and highest in Kallmann's study (cited in Gottesman and Shields 1982), which by far has the largest sample. The older studies have been criticized for selection bias with preferential ascertainment of concordant twins, overweight of severe cases, diagnostic concepts which are too wide, and no blind assessment. Some of this may apply to series based on hospital residence samples, but not to the series of Essen-Möller and Slater (cited in Gottesman and Shields 1982), which were based on consecutive samples. The more recent studies have tried to make up for the shortcomings of the older and have managed to get more unselected and representative series by using twin and case registers. The newer studies found lower monozygotic rates, ranging from .18-.50. The actual sizes of the concordance rates are, however, less interesting than the ratios between the monozygotic and dizygotic rates, which in both the older and the newer studies vary between 2 : 1 and 8 : 1 — all confirming the genetic component in schizophrenia (Tables 1, 2).

The rates and the ratios are both highly dependent on the diagnostic criteria used for selection of probands and evaluation of concordance. Gottesman and Shields (1972) presented their Maudsley twins case histories to a number of clinical assessors of varying diagnostic orientation. A narrow concept based on schneiderian first-rank symptoms produced low rates and failed to show significant differences between monozygotic and dizygotic rates. A broad concept including schizotypal disorders yielded the highest rates, but did not discriminate between the monozygotic and dizygotic rates to the same degree as the more intermediate consensus diagnoses, which produced the highest ratios of 4 or 5 to 1 (Tables 3, 4).

McGuffin and colleagues (1987a) have applied operationalized diagnostic criteria to the Maudsley twins and showed that the strict Research Diagnostic Criteria (RDC) and the DSM-III criteria produced the highest ratios of about 5 to 1, whereas the restrictive schneiderian criteria cut out most of the twins and appeared useless. Diagnostic definition by the RDC or DSM-III criteria for schizophrenia for selection of probands and evaluation of concordance thus appear to be the most promising for molecular research, by identifying diagnostic groupings with high heritability (Table 5).

Breaking schizophrenia down into clinical subtypes like paranoid/nonparanoid, statistically derived cluster types, or the Crow types with positive and negative symptoms does not bring out any marked differences in heritability. McGuffin et al. (1987b) subdiagnosed the Maudsley twins as to these subtypes. The concordances for schizophrenia were higher for nonparanoid, that is, hebephrenic and undifferentiated subtypes, than for the paranoid subtypes, but the ratio was slightly lower for the nonparanoid than the paranoid subtypes. For the statistically derived cluster types the concordances were higher for the H type than for the P type, but again the ratio was lower in the subgroup with the higher concordances. Only the Crow types showed a remarkably high ratio which, however, was for the mixed type. The Crow type II had only few twins, and none of them monozygotic. The homotypic concordances in the schizophrenia concordant twins were quite high in all subtypes, ranging from .73 to 1.00.

As to manifestation rates or penetrance the concordance rates indentical twins may serve as an indicator. Assuming monozygotic concordance to be representative for manifestation in the general population, the rates in the newer series between .20 and .50 give an estimation of a penetrance of 20%-50%, depending on the broadness of concordance accepted.

Expressivity may possibly be inferred from the broad concordance figures including attenuated or less severe forms of schizophrenia as concordant, such as schizophreniform or borderline cases, or even counting schizoid, odd, eccentric, or nervous twins as concordant. In the Danish study (Fischer 1973) the broadest concordance increased the rates to .64 for monozygotic cotwins, and .41 for dizygotic cotwins, but the low ratio indicates that these figures include nongenetic

Table 1. Older schizophrenic twin series, pairwise concordance (references see Gottesman and Shields 1982)

	Monozygotic pairs		Dizygotic pairs (same sexed)	
	n	Concordance	*n*	Concordance
Luxenburger (1928-1936)	19	.58	13	.00
Rosanoff et al. (1934)	41	.61	53	.13
Essen Möller (1941-1970)	11	.64	27	.15
Kallmann (1946)	174	.69	296	.11
Slater (1957)	37	.65	58	.14
Inouye (1961-1972)	58	.59	20	.15

Table 2. Newer schizophrenic twin series, pairwise concordance (updated from Gottesman and Shields 1982)

	Monozygotic pairs		Dizygotic pairs (same sexed)	
	n	Concordance	*n*	Concordance
Tienari (1963-1975)	20	.30	42	.14
Kringlen (1967)	55	.38	90	.10
Fischer (1969)	21	.48	41	.19
Allen et al. (1972)	95	.27	125	.05
Gottesman and Shields (1972)	22	.50	33	.09
Kendler and Robinette (1984)	164	.18	268	.03
Onstad et al. (1991)	24	.33	28	.04

Table 3. Twin concordance in relation to diagnostic orientation (the Maudsley twin series; Gottesman and Shields 1972, 1982)

	Pairwise concordance	
	Monozygotic twins	Dizygotic twins
Narrow (Birley)	.20	.14
Consensus strict	.40	.10
Consensus extended	.50	.09
Broad (Meehl)	.54	.21

Table 4. Twin concordance in schizophrenia by various operationalized criteria (the Maudsley twin series; McGuffin et al. 1987a)

	Probandwise concordance	
	Monozygotic twins	Dizygotic twins
Feighner criteria	.47	.11
RDC	.53	.10
DSM-III	.48	.10
Schneider criteria	.22	.50

Table 5. Twin concordance in schizophrenia subgroups (McGuffin et al. 1987b)

Probandwise concordance for schizophrenia		
Proband	Monozygotic twins	Dizygotic twins
Paranoid	.40	.07
Nonparanoid	.62	.15
Cluster type P	.33	.06
Cluster type H	.79	.18
Crow type I	.53	.19
Crow type II	-	.00
Crow mixed type	.64	.00

Table 6. Schizophrenia and schizophrenialike psychosis in offspring of discordant monozygotic and dizygotic schizophrenic twin pairs

		n	S + S-like	MR (%)
MZ	Probands	14	1	10.0
	Co-twins	24	4	17.4
DZ	Probands	13	1	8.3
	Co-twins	52	1	2.1

Age correction by the Kaplan-Meier Estimate.
S, schizophrenia; MR, morbid risk; MZ, monozygotic; DZ, dizygotic

cases, and therefore should probably not be used for estimation of expressivity.

Phenotypic variability or heterogeneity is also reflected in the variability of disorders included in the broad concordance. Schizoaffective, paranoid, and atypical psychosis are among those most frequently included. It is remarkable that so far no case of typical manic-depressive disorder has been decribed in cotwins of typical schizophrenic proband twins.

The high ratio between monozygotic and dizygotic rates is an indication of genetic heterogeneity. A single major gene would be expected to produce a ratio of 2 : 1. A ratio of 5 : 1 points to polygenetic inheritance, which may appear less encouraging to molecular research. All the same, identification of one of these genes may open the door to understanding pathogenetic mechanisms of major importance for future treatment (Table 6).

Etiological heterogeneity as to nongenetic phenocopies may be estimated by a special form of twin research: The study of the children of discordant identical twins. In the Danish twin study of schizophrenia, Margit Fischer (1973) observed almost equal risks of schizophrenia in the children of the monozogotic twins. A recent 16-year update of the Fischer twin offspring based on register information and case reports (Gottesman and Bertelsen 1989) has confirmed her observations. In this study it was possible to include the offspring of dizygotic twins. The risk in the children of the nonschizophrenic dizygotic twins was much lower, of the same magnitude as in second-degree relatives, specifically, nephews and nieces of the schizophrenic dizygotic twins. This shows that the genotype for schizophrenia can remain completely unexpressed throughout the lifetime of the nonschizophrenic twins, but all the same be transmitted and expressed in the next generation. This finding suggests that phenocopies are infrequent and increases the probability of detecting a major gene locus by molecular research.

Manic-Depressive Disorders

Twin studies of manic-depressive disorders also show higher rates in monozygotic than in dizygotic twins, particularly in some of the older studies based on hospital residence samples and reaching almost full concordance in Kallmann's study, which unfortunately never was well documented by case histories (Table 7).

A Danish psychiatrist working at Kallmann's department at the time of the study has confirmed that Kallmann's diagnoses were quite reliable and not based on a particularly wide definition. The high monozygotic concordance therefore may be explained by the inclusion of a majority of severe bipolar cases. Anyway, twin studies have demonstrated a strong genetic factor in manic-depressive disorders, too.

Some of the more recent studies have tried to identify subgroups with high heritability (Table 8), using the clinical subdivision into bipolar and unipolar disorders. The twin studies confirm the bipolar and unipolar distinction by demonstrating a significantly lower number of concordant monozygotic pairs with one twin bipolar and the other unipolar than expected in case of one illness only. On the other hand, this shows that a bipolar genotype may only expressed as unipolar (Table 9).

Pairs with bipolar disorder show a higher monozygotic concordance than unipolar pairs and the broad bipolar concordance reaches almost 100%. The bipolar distinction, therefore, indentifies a group with particularly high heritability, suggesting a single major locus gene with almost full penetrance, hence encouraging molecular research (Table 10).

The phenotypic variability is reflected by the variety of disorders included in the broad monozygotic concordance, suggesting the range of disorders that should be included as secondary cases in molecular research. In the Danish study (Bertelsen et al. 1977) the variation included schizoaffective psychosis, atypical depressive psychosis, and affective personality disorder of cycloid or hypomanic type, corresponding to the DSM-III-R and ICD-10 cyclothymia. The affective personality disorders may also be considered milder forms of bipolar disorder and thus more reflect expressivity than pleiotropy (Table 11).

The variety of disorders included in broad unipolar monozygotic concordance suggests that besides atypical manifestations and attenuated forms unipolar depression may include both bipolar genotypes with only unipolar manifestation and nongenetic phenocopies.

Table 7. Pairwise concordance in manic-depressive twin series (updated from Bertelsen et al. 1977)

	Monozygotic pairs		Dizygotic pairs	
	n	Concordance	*n*	Concordance
Luxenburger (1928)	3	.67	13	.00
Rosanoff et al. (1934)	23	.70	67	.16
Essen-Möller (1941)	8	.25	3	.00
Slater (1953)	8	.50	30	.24
Kallmann (1954)	27	.93	55	.24
Da Fonseca (1959)	21	.71	39	.38
Kringlen (1967)	6	.33	20	.00
Allen et al. (1974)	15	.33	34	.00
Bertelsen et al. (1977)	55	.58	52	.17
Kringlen (1980)	18	.67	64	.09

Table 8. Bipolar-unipolar concordance in monozygotic twins (Zerbin-Rüdin 1969; Bertelsen et al. 1977)

	Series	Casuistics
Both unipolar	20	13
Both bipolar	18	17
One unipolar, one bipolar	9	5

Table 9. Monozygotic probandwise concordance in bipolar and unipolar affective disorders

		Concordance	
Bipolar	*n*	Strict	Broad
Bertelsen (1977)			
BP-I	25	.80	.96
BP-II	9	.78	1.00
Kringlen (1980)	8	.38	.75
Maudsley (1990)			1.00
(Rifkin and Gurling 1991)			
Unipolar			
Bertelsen (1977)			
≥ 3 episodes	29	.59	.79
< 3 episodes	6	.33	.67
Kringlen (1980)	10	.60	
McGuffin et al..(1991)	62	.53	
(DSM-III major depression)			

Table 10. Bipolar monozygotic broad concordance (six cases)

Bipolar-I	Atypical psychosis (schizoaffective)
	Atypical psychosis (delirious and depressive symptoms)
	Melancholiform episodes in symptomatic epilepsia
	Affective personality disorder (cycloid)
	Affective personality disorder (cycloid)
Bipolar-II	Affective personality disorder (hypomanic), with reactive depression

Table 11. Unipolar monozygotic broad concordance (eight cases)

≥ 3 Episodes	Atypical psychosis (schizoaffective)
	Atypical psychosis (depression with delirious traits)
	Psychosis, acute hallucinatory —> death
	Senile depression —> dementia
	Affective personality disorder (cycloid)
	Affective personality disorder (depressive)
< 3 Episodes	Senile psychosis (depressive and paranoid)
	Personality disorder, affective?, suicide

Table 12. Schizoaffective psychoses: twin studies

		Pair (*n*)	Pairwise concordance for endogenous psychosis
US Study (Cohen et al. 1972)	MZ	14	.50
	DZ	12	.00
Danish studies (Fischer 1973; Bertelsen 1977)	MZ	7	.71
	DZ	7	.00

MZ, monozygotic; DZ, dizygotic

The frequency of nongenetic phenocopies may again be estimated in twin offspring studies. The Danish study of children of twin pairs discordant for manic-depressive disorder mainly include unipolar pairs, because the majority of the bipolar cases are concordant.

The risk of manic-depressive disorder in the offspring of monozygotic discordant cotwins is 16.7%. This is significantly higher than the risk in children of discordant dizygotic cotwins of 1.6%, which is of the same magnitude as in second-degree relatives. This confirms the existence of an unexpressed genotype which is transmitted to the next generation, indicating a low frequency of phenocopies, a fact which should encourage molecular research of unipolar depression, too.

Schizoaffective Psychosis

For schizoaffective psychosis, twin studies are few, and they antedate the introduction of more precise diagnostic criteria. The American study and the extracted data from the Danish studies are based on ICD-8 diagnostic guidelines (Table 12). They both show remarkably high ratios between monozygotic and dizygotic rates. In the American study (Cohen et al. 1972) the monozygotic rate is higher than the corresponding rates for schizophrenic and manic-depressive twin pairs from the same study population. Also the data from the Danish studies (Fischer 1973; Bertelsen et al. 1977) show a high monozygotic rate of .71, or even 1.00 if suspected cases are included, compared to .48 for the schizophrenia study and .58 for the manic-depressive study. In the Danish series the concordant cotwins were homotypic in only one third of the cases, otherwise they had affective psychosis or schizophrenia. The dizygotic rate in both schizoaffective series is zero, which also appears remarkable. The findings suggest a disorder with a very strong genetic component. The high monozygotic concordance rate may, however, be explained by a mixed affective and schizophrenic inheritance. The additive or interactive effect of several different loci may produce an atypical manifestation with high penetrance and a high concordance rate in monozygotic twins, who are identical in all loci. The chance for dizygotic twins of being identical in one locus is one half for each locus, and in the case of several loci, the probability of identity in all loci will be very low, giving a low dizygotic concordance and a high ratio between monozygotic and dizygotic rates. Suggestions that schizoaffective psychosis presents an intermediate form of a continuum psychosis with manic-depressive disorder and schizophrenia at the extreme ends of the continuum are contradicted by the lack of evidence in monozygotic twin paris with one twin with typical schizophrenia and the other with typical manic-depressive disorder. Recent publications (Fischer 1980; McGuffin et al. 1982; Dalby et al. 1986) have reported the observation of what appeared to be such combinations in identical twins or triplets, which, however, on close inspection have shown one partner to be schizoaffective as a phenotypic variation of schizophrenia or manic-depressive disorder. Recent studies in molecular genetics have included schizoaffective disorders among affected relatives for linkage evaluation in both schizophrenia and manic-depressive studies. Twin research supports this notion, and so far cannot provide a basis for molecular research on schizoaffective disorder as such.

Conclusion

The importance of twin research in the era of molecular genetics is demonstrated by evidence from twin studies on the major psychoses. Twin research contributes to molecular genetics by demonstrating a genetic factor, by identifying well-defined diagnostic groups or subgroups with particularly high heritability, by evaluating penetrance and expressivity, by assessing phenotypic variation of disorders to be included as secondary cases in linkage studies, and by estimating the frequency of phenocopies through the demonstration of equal risks in twin-offspring studies of discordant twin pairs. These contributions apply equally well to other psychiatric disorders such as obsessive-compulsive disorder, eating disorders, and childhood autism. Providing guidelines for molecular research, twin research still has a significant role to play in psychiatric genetics which is of major importance for future classification and treatment of psychiatric disorders.

References

Bertelsen A, Harvald B, Hauge M (1977) A Danish twin study of manic-depressive disorders. Br J Psychiatry 130:330-335

Cohen SM, Allen MG, Pollin W, Hrubec Z (1972) Relationship of schizoaffective psychosis to manic-depressive psychosis and schizophrenia. Arch Gen Psychiatry 26:539-546

Dalby T, Morgan D, Lee ML (1986) Single case study, schizophrenia and mania in identical twin brothers. J Nerv Ment Dis 174:304-308

Fischer M (1973) Genetic and environmental factors in schizophrenia, a study of schizophrenic twins and their families. Munksgard, Copenhagen

Fischer M (1980) Twin studies and dual mating studies in defining mania. In: Belmaker RH, van Praag HM (eds) Mania, an evolving concept. Spectrum, Jamaica NY, pp 43-60

Gottesman II, Bertelsen A (1989) Confirming unexpressed genotypes for schizophrenia. Arch Gen Psychiatry 46:867-872

Gottesman II, Shields J (1972) Schizophrenia and genetics, a twin study vantage point. Academic, New York

Gottesman II, Shields J (1982) Schizophrenia, the epigenetic puzzle. Cambridge University Press, Cambridge

Kringlen E (1980) On the genetics and nosology of affective disorder. In: Achté K, Aalberg V, Lönngvist J (eds) Psychopathology of depression. Psychiatr Fennica [Suppl] 1980:19-25

McGuffin P, Reveley A, Holland A (1982) Identical triplets: nonidentical psychosis? Br J Psychiatry 140: 1-6

McGuffin P, Farmer A, Gottesman II (1987a) Modern diagnostic criteria and genetic studies of schizophrenia. In: Häfner H, Gattaz WF, Janzarik W (eds) Search for the causes of schizophrenia. Springer, Berlin Heidelberg New York, pp 143-156

McGuffin P, Farmer A, Gottesman II (1987b) Is there really a split in schizophrenia, the genetic evidence. Br J Psychiatry 150:581-592

McGuffin P, Katz R, Rutherford J (1991) Nature, nuture and depression: a twin study. Psychol Med 21:329-335

Onstad S, Skre I, Torgersen S, Kringlen E (1991) Twin concordance for DSM-III-R schizophrenia. Acta Psychiat Scand 83:395-401

Rifkin L, Gurling H (1991) Genetic aspects of affective disorders. In: Horton RW, Katona CLL (eds) Biological aspects of affective disorders. Academic, London, p 313

Zerbin-Rüdin E (1969) Zur Genetik der depressiven Erkrankungen. In: Hippius H, Selbach H (eds) Das depressive Syndrom. Urban und Schwartzenberg, Berlin, pp 37-56

Discussion of Presentation A. Bertelsen

Tsuang
In the schizophrenia research, where would you like to put the schizoaffective disorders from the twin studies. Do you have any advice for the selection of the probands?

Bertelsen
Not based on these studies. I am not able to conclude that depressive forms or manic forms go more with schizophrenia or manic-depressive psychosis. From twin research, it appears so far that they are evenly distributed, but the data are still inconclusive.

McGuffin
I think there is a slight hint from the twin research, only a slight hint because the numbers are small. But in another reanalysis from the Gottesman and Shields Maudsley data, which actually you did not show, we tried playing around with diagnostic boundaries. And there, the degree of genetic determination as reflected in the MC to DC ratio — if you feel that is reasonable to do — actually increased when we added so-called affective psychoses with mood-incongruent delusions. I have a rather strong hunch that whether they are schizomanic or schizodepressed, people with mood-incongruent delusions fall on the schizophrenia side rather than on the affective disorder side.

Mendlewicz
You said that there are no reports of monozygotic twins where one is schizophrenic and the other one manic-depressive.

Bertelsen
Not typical schizophrenic and typical manic-depressive. That is right.

Blackwood
It worries me that some of the patients in the examples you gave were rediagnosed afterwards to be schizoaffective rather than schizophrenic. I remember one example in our hospital of schizophrenia and manic depression, but one could not publish just a single case. How confident are you that there is no separation of the two. You showed four published cases, showing different psychoses, and I am wondering if

they have been rediagnosed as being schizoaffective rather than schizo-phrenic. Were you yourself rediagnosing?

Bertelsen
They published extensive case histories, and from these case histories it appeared that they had marked affective or schizophrenic features, but were not either typical schizophrenic or manic-depressive. Schizoaffective is used here in a broad ICD-8 unterstanding.

McGuffin
I broadly agree with Dr. Bertelsen's reading of the literature. In the set of triplets we reported on, we prepared detailed abstracts and distributed these to three well-known British psychiatrists, along with a lot of dummy cases which had nothing to do with the triplets, to hear what their clinical diagnosis was. We also did two separate sets of standarized interviews — PSE and SADS — and the results were a mess. It really was a matter of where to draw the lines. The clinical diagnoses were that two of the triplets had schizophrenia and one had manic-depressive disorder. The Maudsley diagnosis on the one who is called manic-depressive is still called manic-depressive disorder. But, he was a manic-depressive with mood-incongruent delusions, so I would tend to agree with Bertelsen's interpretation.

Gurling
Dr. Bertelsen, I think in dual mating studies you also find independent segregation of manic-depressive psychosis and schizophrenia.

Bertelsen
Yes, the offspring of combinations with one parent being schizophrenic and the other manic-depressive do not have schizoaffective children, but rather children with schizophrenic or affective, mainly affective, psychoses. And only a few were schizo-affective.

Assessing the Evidence for Linkage in Psychiatric Genetics*

J. Ott and J. D. Terwilliger

Introduction

Linkage analysis refers to that branch of genetics in which the recombination fraction θ between two loci is estimated, where $\theta = 1/2$ represents free recombination (absence of linkage) and $\theta < 1/2$ refers to linkage. In human linkage analysis, the evidence for linkage between a trait and a marker (or a map of markers) is generally summarized by the maximum logarithm of odds (lod) score, Z_{max}. Conventionally, if Z_{max} attains or exceeds the value 3, linkage is said to be "proven." Such a "proof," however, is of a statistical nature — a Z_{max} value of 3 or more may occur by chance (without linkage) although this is very unlikely. Given that no linkage exists, the probability (the so-called *p* value) of finding a maximum lod score equal to or exceeding Z is no larger than 10^{-Z} (Morton 1978; see also discussion in Ott 1991).

Because of the low prior probability (~ 0.02) that two loci are within measurable distance of each other, with an observed maximum lod score $Z_{max} = 3$, the posterior odds for linkage are equal to approximately 20 : 1 only (prior odds = 1 : 50, observed odds = 1000 : 1). With $Z_{max} = 5$, on the other hand, the posterior odds for linkage are increased to 2000 : 1. Statistically, maximum lod scores on the order of 5 or higher are an almost certain indication of linkage and should practically never occur except when there is linkage. However, in psychiatric genetics, some impressive looking results have been found to vanish upon follow-up. The most spectacular case is that of autosomal affective disorder for which linkage to markers on chromosome 11 was reported (Egeland et al. 1987). Upon reevaluation of disease and marker phenotypes and expansion of the pedigree to include additional branches, however, linkage had to be excluded for that chromosome 11 segment (Kelsoe et al. 1989). In another case, the situation is seemingly still unresolved. While a linkage analysis between schizophrenia and chromosome 5 markers resulted in lod scores exceeding 6 (Sherrington et al. 1988), other researchers have been unable to replicate this result (for example, Kennedy et al. 1988; Kaufmann et al. 1989; Aschauer et al. 1990) and the existence of a gene for schizophrenia on chromosome 5 now appears rather doubtful. What are the reasons for the apparent inflation of maximum lod scores in psychiatric genetics?

*This work was supported by grant MH44292 from the National Institutes of Mental Health.

Causes of Lod Score Inflation in Psychiatric Genetics

One of the reasons for an increased rate of false-positive results might be that researchers did not make allowance for the number of markers tested. In classical linkage analyses with known monogenic disease, such adjustments are not necessary, because with an increased number of markers, the elevated prior probability of linkage more than offsets the increased type I error rate (Ott 1991). However, in complex diseases such as schizophrenia in which the mode of inheritance is not well defined and involvement of a single major gene has not clearly been demonstrated, one must adjust the critical lod score for the number *n* of unlinked markers tested, for example, by increasing the critical level of $Z_{max} = 3$ by $\log_{10}(n)$ (Kidd and Ott 1984).

In linkage analysis of schizophrenia and other psychiatric traits, a major problem is that the mode of inheritance has not been well established. A common strategy is to try different modes of inheritance, different values for the penetrance, and several classification schemes and to see which set of model parameters yields the highest lod score. The original idea behind this strategy was that it would lead both to the detection of linkage and the delineation of the mode of inheritance. It has been pointed out, however, that this approach is based on circular reasoning and might lead to an inflation of the maximum lod score (Ott 1990a). Earlier, lod score inflation was not believed to occur, on the grounds that analysis under wrong model parameters does not tend to increase the expected lod score (Clerget-Darpoux et al. 1986; Greenberg 1989). However, the practice of maximizing the lod score over models is statistically different from analyzing data under a wrong model and has been demonstrated to inflate the maximum lod score (Clerget-Darpoux et al. 1990; Weeks et al. 1990a). Below, ways of assessing the significance of an observed maximum lod score (even when maximized over models) are outlined.

Other reasons for lod score inflation may be of a psychological nature. For example, when the person determining affection status is aware of the marker genotypes, or when the molecular geneticist knows which family member is affected or unaffected, biases are almost inevitable. One then tends to scrutinize apparent recombinants more rigorously than nonrecombinants which may bias the results in favor of nonrecombinants and, thus, a higher lod score. This problem is particularly relevant in psychiatric genetics when diagnoses are relatively soft.

Evaluating Maximum Lod Scores by Computer Simulation

To evaluate a lod score which has been maximized over various models, one approach is to simulate the maximization procedure by computer and to carry out this maximization under the assumption of no linkage (true recombination fraction, $r = 1/2$). Any maximum lod score is then known to be due to causes other than linkage. In practice, one assumes family data and marker loci like those used in the original investigation. Rather than working with the marker genotypes originally observed, one randomly generates marker genotypes on the computer according to population genetics principles and the mendelian laws (Weeks et al. 1990a). This is done for all individuals for whom marker genotypes had been observed in the original investigation. The generated marker genotypes, along with the known trait phenotypes, are then analyzed in the same manner as in the original investigation, that is, with maximization over models. This results in a generated maximum lod score and completes one round (replicate) of the computer simulation. The whole process is repeated and leads to another generated maximum lod score. Eventually, from a large number of such replicates one determines the proportion (p) of replicates in which the generated maximum lod score was greater than or equal to the one observed in the original investigation. This proportion represents an estimate of the p value associated with the observed maximum lod score.

Simulating marker genotypes with linkage to the given trait is complicated and is generally used for power calculations. It is usually carried out with the aid of computer programs such as SIMLINK (Ploughman and Boehnke 1989) and SLINK (Weeks et al. 1990b). Because of their complexity and generality, however, these programs are rather slow. Under absence of linkage ($r = 1/2$), the marker genotypes are independent of the trait phenotypes so that much simpler and faster procedures than those implemented in the general simulation programs may be employed. We have incorporated such a procedure in a computer program, MOM (for Maximization Over Models). It is written in Pascal for Vax computers running under VMS and requires the LINKMAP program of the LINKAGE package for the calculation of lod scores in each replicate. It is freely available from us.

The proportion p of replicates with a maximum lod score equal to or exceeding the one originally observed is an estimate of the true p value associated with the observed maximum lod score. Because of sampling fluctuations, one should calculate a confidence interval for p, for example, at the 95% confidence level, which may be done with the BINOM program (one of our Linkage Utility programs). To declare an observed maximum lod score significant in the classical sense, the upper bound of the confidence interval (not just p) should be no larger than 0.001. In practice, this means that for sufficient accuracy of the p value, the number of replicates must be at least 3000 (Ott 1991).

Importance Sampling

Under absence of linkage, few generated maximum lod scores will exceed the observed maximum lod score so that the p value will not be estimated very accurately. Instead of simple random sampling as described above for the computer simulation, one may employ a technique called importance sampling, which allows one to concentrate the distribution of the sample points in the parts of the interval that are of most "importance" (Hammersley and Handscomb 1983). For the simulations described above, with importance sampling, marker genotypes are generated under linkage ($r < 1/2$) so that large generated maximum lod scores occur more frequently. A correction factor is then applied to the generated maximum lod scores to reduce the weight of large values. Under importance sampling, only a small number of replicates is required to furnish an accurate estimate of the p value. However, this advantage is purchased at a cost — because importance sampling is carried out under linkage, marker genotypes are no longer independent of the trait phenotypes. Consequently, one must specify a mode of inheritance for the trait, which is not required for simulations under free recombination. Also, one can no longer use the MOM program but must resort to the slower programs referred to above. We are presently investigating properties of importance sampling and possible applications of it for computer simulation in human linkage analysis.

Formally, importance sampling may be characterized as follows. As outlined by Boehnke (1986), one generally samples from the conditional distribution, $P(g|x; r)$, where $g = (g_1..g_m)$ is an array of marker genotypes (g_m = genotype of mth family member) and x is the array of observed trait phenotypes. In a computer simulation, one approximates the distribution $P(g|x; r)$ by a random sample from it, where each sampled genotype array has equal probability of occurrence in the sample. For the determination of p values, with simple random sampling, one samples from $P(g|x; 1/2)$ and calculates a value of Z_{max} for each g, where the resulting distribution of the Z_{max} has a thin upper tail. To obtain a thicker upper tail of Z_{max} values, one samples from $P(g|x; r < 1/2)$, that is, one samples under linkage and adjusts each generated Z_{max} by a correction factor, $P(g|x; 1/2)/P(g|x; r) = L(1/2)/L(r) = 10^{-Z}$, where $L(r)$ is simply the pedigree likelihood for trait phenotypes x and sampled marker genotypes g and may be calculated in a linkage program; Z is the lod score obtained in the given replicate at the generating value of r.

Discussion

The methods discussed above allow one to interpret a maximized lod score in terms of a p value, provided that the various ways in which the lod score was maximized in the original investigation are known. If some of the maximization steps were not reported they cannot be taken into account in the simulation so that some inflation of the lod score is expected to remain.

As one can correct for inflation of the lod score as described above, is it recommendable to search for linkage over a large number of different models? The answer to this question is not clear; such a strategy might have relatively low power, but this has not been investigated. It appears preferable to limit the analyses to a small number of models, for example, dominant and recessive inheritance, each with medium penetrance. Also, two-locus models (Lathrop and Ott 1990) seem to be more realistic than monogenic inheritance. With regards to classification schemes, when phenotypes can be represented in ordered classes of certainty ranging from clearly affected through clearly unaffected, it may be advantageous to combine all of them in a single analysis in which the classes of certainty are characterized by different penetrances (Ott 1990b).

References

Aschauer HN, Aschauer-Treiber G, Isenberg KE, Todd RD, Knesevich MA, Garver DL, Reich T, Cloninger CR (1990) No evidence for linkage between chromosome 5 markers and schizophrenia. Hum Hered 40:109-115

Boehnke M (1986) Estimating the power of a proposed linkage study: a practical computer simulation approach. Am J Hum Genet 39:513-527

Clerget-Darpoux F, Bonaïti-Pellié C, Hochez J (1986) Effects of misspecifying genetic parameters in lod score analysis. Biometrics 42:393-399

Clerget-Darpoux F, Babron M-C, Bonaïti-Pellié C (1990) Assessing the effect of multiple linkage tests in complex diseases. Genet Epidemiol 7:245-253

Egeland JA, Gerhard DS, Pauls D, Sussex JN, Kidd KK, Allen CR, Hostetter AM, Housman DE (1987) Bipolar affective disorders linked to DNA markers on chromosome 11. Nature 325:783-787

Greenberg DA (1989) Inferring mode of inheritance by comparison of lod scores. Am J Med Genet 34:480-486

Hammersley JM, Handscomb DC (1983) Monte Carlo methods. Chapman and Hall, New York

Kaufmann CA, DeLisi LE, Lehner T, Gilliam TC (1989) Physical mapping, linkage analysis of a putative schizophrenia locus on chromosome 5q. Schizophr Bull 15:441-452

Kelsoe JR, Ginns EI, Egeland JA, Gerhard DS, Goldstein AM, Bale SJ, Pauls DL, Long RT, Kidd KK, Conte G, Housman DE, Paul SM (1989) Reevaluation of the linkage relationship between chromosome 11p loci and the gene for bipolar affective disorder in the Old Order Amish. Nature 342:238-243

Kennedy JL, Giuffra LA, Moises HW, Cavalli-Sforza LL, Pakstis AJ, Kidd JR, Castiglione CM, Sjögren B, Wetterberg L, Kidd KK (1988) Evidence against linkage of schizophrenia to markers on chromosome 5 in a northern Swedish pedigree. Nature 336:167-170

Kidd KK, Ott J (1984) Power and sample size in linkage studies. Cytogenet Cell Genet 37:510-11

Lathrop GM, Ott J (1990) Analysis of complex diseases under oligogenic models and intrafamilial heterogeneity by the LINKAGE programs. Am J Hum Genet 47:A188 (Abstr)

Morton NE (1978) Analysis of crossingover in man. Cytogenet Cell Genet 22:15-36

Ott J (1990a) Genetic linkage and complex diseases: a comment. Genet Epidemiol 7:35-36

Ott J (1990b) Genetic linkage analysis under uncertain disease definition. In: Cloninger CR, Begleiter H (eds) Genetics and biology and alcoholism. Cold Spring Harbor Laboratory Press, Cold Spring Harbor, pp 327-331 (Banbury Report 33)

Ott J (1991) Analysis of human genetic linkage, 2nd edn. Johns Hopkins University Press, Baltimore

Ploughman LM, Boehnke M (1989) Estimating the power of a proposed linkage study for a complex genetic trait. Am J Hum Genet 44:543-551

Sherrington R, Brynjolfsson J, Petursson H, Potter M, Dudleston K, Barraclough B, Wasmuth J, Dobbs M, Gurling H (1988) Localization of a susceptibility locus for schizophrenia on chromosome 5. Nature 336:164-167

Weeks DE, Lehner T, Squires-Wheeler E, Kaufmann C, Ott J (1990a) Measuring the inflation of the lod score due to its maximization over model parameter values in human linkage analysis. Genet Epidemiol 7:237-243

Weeks DE, Ott J, Lathrop GM (1990b) SLINK: a general simulation program for linkage analysis. Am J Hum Genet 47:A204 (Abstr)

Editors:

Professor Dr. Julien Mendlewicz
University Hospital of Brussels
Hôpital Erasme
Route de Lennik 808
B-1070 Bruxelles
Belgium

Professor Dr. Hanns Hippius
Head of Dept. of Psychiatry
University of Munich
Nußbaumstraße 7
D-W-8000 München 2

Co-Editors:

PD Dr. Brigitta Bondy
Dept. of Psychiatry
University of Munich
Nußbaumstraße 7
D-W-8000 München 2
Germany

Professor Dr. Manfred Ackenheil
Dept. of Psychiatry
University of Munich
Nußbaumstraße 7
D-W-8000 München 2
Germany

Professor Dr. Merton Sandler
Dept. of Chemical Pathology
Queen Charlotte's Maternity Hospital
Goldhawk Road, London W6 OXG
United Kingdom

Workshop Participants and Senior Authors

Professor Dr. Manfred Ackenheil
Dept. of Psychiatry
University of Munich
Nußbaumstraße 7
D-W-8000 München 2
Germany

Professor Dr. Luigi A. Amaducci
Dept. of Neurology
University of Florence
Viale Morgagni 85
I-50134 Florence
Italy

Professor Dr. Eric Barnard
Director of MRC Molecular
Neurobiology Unit
Medical Research Council Centre
Hills Road
Cambridge CB2 2QH
United Kingdom

Miron Baron, M. D.
Professor and Director
Division of Psychogenetics
Dept. of Medical Genetics
New York State Psychiatric Institute
and Dept. of Psychiatry, Columbia
University College of Physicians & Surgeons
722 West 168th Street
New York, NY 10032
USA

Dr. Aksel Bertelsen
Institute of Psychiatric Demography
Aarhus University Psychiatric Hospital
DK - 8240 Risskov
Denmark

Professor Dr. Konrad Beyreuther
Laboratory for Molecular
Neuropathologie
University of Heidelberg
Im Neuenheimer Feld 282
D-W-6900 Heidelberg
Germany

Dr. Douglas Blackwood
Department of Psychiatry
The University of Edinburgh
Kennedy Tower
Royal Edinburgh Hospital
Morningside Park
Edinburgh EH10 5HF
United Kingdom

PD Dr. Brigitta Bondy
Dept. of Psychiatry
University of Munich
Nußbaumstraße 7
D-W-8000 München 2
Germany

Dr. Christina Van Broeckhoven
Neurogenetics Laboratory
Born Bunge Foundation
University of Antwerp (UTA)
Dept. of Biochemistry
Universiteitsplein 1
B-2610 Antwerpen
Belgium

Marc G. Caron, Ph. D.
Professor of Cell Biology
Duke University Medical Center
Dept. of Cell Biology
Box 3287
Durham, NC 27710
USA

C. Robert Cloninger, M. D.
Wallace Renard Professor
and Head Dept. of Psychiatry
Washington University
School of Medicine
Medical School Box 8134
4940 Audubon Avenue
St. Louis, MO 63110
USA

Dr. Alec Coppen
Ashtead Hospital
The Warren
Ashtead, Surrey KT21 2SN
United Kingdom

Dr. Timothy J. Crow
Division of Psychiatry
Clinical Research Centre
Watford Road
Harrow, Middlesex HA1 3UJ
United Kingdom

Professor Dr. Rolf R. Engel
Dept. of Psychiatry
University of Munich
Nußbaumstrasse 7
D-W-8000 München 2
Germany

PD Dr. Ursula Froster-Iskenius
Dept. of Gynecology and Obstetrics
University of Lübeck
Ratzeburger Allee 160
D-W-2400 Lübeck
Germany

Dr. Hugh Gurling
Department of Psychiatry
University College and
Middlesex School of Medicine
Riding House Street
London W1P 7PN
United Kingdom

PD Dr. Johannes Hebebrand
Dept. of Child and Adolescent
Psychiatry
University of Marburg
Hans-Sachs-Straße 6
D-W-3550 Marburg
Germany

Professor Dr. Hanns Hippius
Head of Dept. of Psychiatry
University of Munich
Nußbaumstraße 7
D-W-8000 München 2
Germany

Dr. Norbert Kathmann
Dept. of Psychiatry
University of Munich
Nußbaumstraße 7
D-W-8000 München 2
Germany

Kenneth K. Kidd, Ph. D.
Professor of Human Genetics,
Psychiatry, and Biology
Yale University,
Department of Human Genetics
School of Medicine
333 Cedar Street, P. O. Box 3333
New Haven, CT 06510-8005
USA

Dr. Sabine Kussmann
Deutsche Forschungsgemeinschaft
P. O. B. 20 50 04
D-W-5300 Bonn 2
Germany

PD Dr. Wolfgang Maier
Dept. of Psychiatry
University of Mainz
P. O. B. 3960
D-W-6500 Mainz
Germany

Dr. Laurent Mâtre
CIBA-GEIGY AG
Dept.: PH 2.212
Location: K 125.15.01
CH-4002 Basel
Switzerland

Dr. Jacques Mallet
CNRS Laboratory of Cellular
and Molecular Neurobiology
1 ave de la Terrasse
F-91198 Gif-sur-Yvette Cedex
France

Professor Dr. Peter McGuffin
University of Wales
College of Medicine
Dept. of Psychological Medicine
Heath Park
Cardiff CF4 4XN
United Kingdom

Professor Dr. Julien Mendlewicz
University Hospital of Brussels
Hôpital Erasme
Route de Lennik 808
B-1070 Bruxelles
Belgium

Kathleen R. Merikangas, Ph. D.
Associate Professor
of Psychiatry & Epidemiology
Yale University of Medicine
40 Temple Street
New Haven, CT 065610-3223
USA

Katsuhiko Mikoshiba, M. D.
Professor
Institute for Protein Research
Osaka University
3-2 Yamadaoka, Suita
Osaka 565
Japan

Dr. Markus Nöthen
Institute for Human Genetics
University of Bonn
Wilhelmstraße 31
D-W-5300 Bonn 1
Germany

Jürg Ott, Ph. D.
Professor
College of Physicians & Surgeons
of Columbia University
Department of Genetics and Development
Box 58
722 West 168th Street
New York, NY 10032
USA

Dr. Leena Peltonen
Head of the Laboratory
of Molecular Genetics
National Public Health Institute
Mannerheimintie 166
SF-00300 Helsinki
Finland

Dr. Annemarie Poustka
German Cancer Research Center
Im Neuenheimer Feld 506
D-W-6900 Heidelberg
Germany

Professor Dr. Peter Propping
Institute for Human Genetics
University of Bonn
Wilhelmstr. 31
D-W-5300 Bonn 1
Germany

Professor Dr. Giorgo Racagni
Center of Neuropharmacology
Institute of Pharmacological Sciences
University of Milan
Via Balzaretti 9
I-20133 Milano
Italy

Theodore Reich, M. D.
Professor of Psychiatry and Genetics
Washington University School of Medicine
The Jewish Hospital of St. Louis
216 South Kingshighway Boulevard
P. O. Box 14109
St. Louis, MO 63178
USA

Dr. Ulrike Reuner
Medical Academy "Carl Gustav Carus"
Fetscherstraße 74
D-O-8019 Dresden
Germany

Professor Dr. Merton Sandler
Department of Chemical Pathology
Queen Charlotte's Maternity Hospital
Goldhawk Road
London W6 OXG
United Kindom

Dr. Pierre Sokoloff
Unit of Neurobiology and Pharmacology
Centre Paul Broca de l'INSERM
2ter rue d'Alesia
F-75014 Paris
France

S. Sorbi, M.D.
Department of Neurologic and Psychiatric Sciences
University of Florence
Viale G. B. Morgagni 85
I-50134 Florence
Italy

Ming T. Tsuang, M. D., Ph. D., D. Sc.
Professor of Psychiatry and
Director of Psychiatric Epidemiology
Harvard University
Chief, Psychiatric Service
Brockton/West Roxbury VA Medical Center
940 Belmont Street
Brockton, MA 02401
USA

Professor Dr. Dr. h. c. Friedrich Vogel
Professor of Human Genetics
and Head of Department
Institute of Human Genetics
and Anthropology
University of Heidelberg
Im Neuenheimer Feld 328
D-W-6900 Heidelberg
Germany

Dr. Dieter B. Wildenauer
Dept. of Psychiatry
University of Munich
Nußbaumstraße 7
D-W-8000 München 2
Germany

Coauthors

M. Abbar
Service de Psychologie Médicale et de Psychiatrie
CHU Lapeyronie
34059 Montpellier
France

M. Ackenheil
Psychiatrische Klinik der Universität
Nußbaumstraße 7
D-W-München 2
Germany

S. Amadéo
Service de Psychiatrie Adulte
CHU 44220 Nantes
France

T. d'Amato
Service de Psychiatrie Adulte
Hôpital du Vinatler
69677 Lyon-Bron
France

M.-C. Babron
Laboratoire de Génétique Epidémiologique
INSERM U155
Chateau de Longchamps
75016 Paris
France

M. L. Bouthenet
Laboratoire de Physiologie
Faculté de Pharmacie
4 Avenue de l'Oberservatoire
75006 Paris
Franc

D. Campion
Secteur de Santé Mentale
CHS 76300 Sotteville-les-Rouen
France

O. Canseil
Service de Psychiatrie Adulte
Hôpital Sainte-Anne
75014 Paris
France

D. Castelnau
Service de Psychologie Médicale et de Psychiatrie
CHU Lapeyronie
34059 Montpellier
France

F. Clerget-Darpoux
Laboratoire de Génétique Epidémiologique
INSERM U155
Chateau de Longchamps
75016 Paris
France

A. DesLauriers
Service de Psychiatrie Adulte
CHU La Salpêtrière
75013 Paris
France

M. Ertl
Psychiatrische Klinik der Universität
Nußbaumstraße 7
D-W-8000 München 2
Germany

S. V. Faraone
Dept. of Psychiatry and Dept. of Epidemiology
Harvard University
Brockton/West Roxbury
V. A. Medical Center
940 Belmont Street
Brockton, MA 02401
USA

P. Forleo
Dept. of Neurology
University of Florence
Viale Morgagni 85
50134 Florence
Italy

F. Gheysen
Service de Psychiatrie Adulte
CHU Côte-de-Nacre
14000 Caen
France

H. Giedke
Psychiatrische Klinik der Universität
Osianderstraße 22
D-W-7400 Tübingen
Germany

M. Gill
Institute of Psychiatry
De Crespgny Park
London SE 5 8AF
United Kingdom

B. Giros
Unité de Neurobiologie et Pharmacologie
(U. 109) de l'INSERM
Centre Paul Broca
2ter rue d'Alésia
75014 Paris
France

B. Granger
Service de Psychiatrie Adulte
CHU Necker
75015 Paris
France

B. Henriksson
Laboratoire de Neurobiologie
Cellulaire et Moléculaire
CNRS
91198 Gif-sur-Yvette
France

C. Hilbich
Center for Molecular Biology
University of Heidelberg
Im Neuenheimer Feld 282
D-W-6900 Heidelberg
Germany

P. Kioschis
Deutsches Krebsforschungszentrum
Im Neuenheimer Feld 280
D-W-6900 Heidelberg
Germany

G. König
Center for Molecular Biology
University of Heidelberg
Im Neuenheimer Feld 282
D-W-6900 Heidelberg
Germany

P. König
Landesnervenkrankenhaus Valduna
A-6830 Rankweil/Vaduna
Austria

L. Lannfelt
Dept. of Psychiatry and Psychology
Karolinska Hospital
10401 Stockholm
Sweden

M. Leboyer
Laboratoire de Génétique Epidémiologique
INSERM U155
Chateau de Longchamps
75016 Paris
France

H. Loo
Service de Psychiatrie Adulte
Hôpital Sainte-Anne
75014 Paris
France

M. J. Lyons
Dept. of Psychology
Boston University
64 Cummington Street
Boston, MA 02215
USA

A. Malafosse
Laboratoire de Neurobiologie
Cellulaire et Moléculaire
CNRS 91198 Gif-sur-Yvette
France

J.-J. Martin
Laboratory of Neuropathology
Born Bunge Foundation
University of Antwerp (UIA)
Universiteitsplein 1
2610 Antwerp
Belgium

M. P. Martres
Unité de Neurobiologie et Pharmacologie
(U. 109) de l'INSERM
Centre Paul Broca
2ter rue d'Alésia
75014 Paris
France

C. L. Masters
Dept. of Pathology
The University of Melbourne
and The Mental Health
Research Institute of Victoria
Parkville
Victoria 3052
Australia

K. R. Merikangas
Dept. of Psychiatry
Yale Medical School
New Haven, CT 06520
USA

W. J. Muir
Dept. of Psychiatry
The University of Edinburgh
Kennedy Tower
Royal Edinburgh Hospital
Morningside Park
Edinburgh EH10 5HF
United Kingdom

G. Multhaup
Center for Molecular Biology
University of Heidelberg
Im Neuenheimer Feld 282
D-W-6900 Heidelberg
Germany

C. Mundt
Psychiatrische Klinik der Universität
Voß-Straße 4
D-W-6900 Heidelberg
Germany

M. Owen
Institute of Medical Genetics
University of Wales
College of Medicine
Cardiff CF4 4XN
United Kingdom

P. Piersanti
Dept. of Neurology
University of Florence
Viale Morgagni 85
50134 Florence
Italy

M.-F. Poirier
Service de Psychiatrie Adulte
Hôpital Sainte-Anne
75014 Paris
France

O. Sabaté
Laboratoire de Neurobiologie
Cellulaire et Moléculaire
CNRS 91198 Gif-sur-Yvette
France

D. Samolyk
Laboratoire de Neurobiologie
Cellulaire et Moléculaire
CNRS 91198 Gif-sur-Yvette
France

H. Sauer
Psychiatrische Klinik der Universität
Voß-Straße 4
D-W-6900 Heidelberg
Germany

G. Schleuning
Bezirkskrankenhaus Haar
Postfach 11 11
D-W-8013 Haar/München
Germany

J. C. Schwartz
Unité de Neurobiologie et Pharmacologie
(U. 109) de l'INSERM
Centre Paul Broca
2ter rue d'Alésia
75014 Paris
France

S. Sorbi
Dept. of Neurology
University of Florence
Viale Morgagni 85
50134 Florence
Italy

J. D. Terwilliger
Dept. of Genetics and Development
Columbia University
722 West 168th Street
New York, NY 10032
USA

E. Zarifian
Service de Psychiatrie Adulte
CHU Côte-de-Nacre
14000 Caen
France

Acknowledgements

When a workshop report appears in print only five months after the meeting words of special thanks are in order!

Thanks to collaboration among the editors — in particular Dr. Brigitta Bondy — the Congress Project Management GmbH (CPM), Frankfurt am Main, and Springer-Verlag, Heidelberg, the goal was achieved of providing in book form the papers and the ensuing discussions which were presented at the Second Collegium Internationale Neuropsychopharmalogicum (C.I.N.P.) President's Workshop to C.I.N.P. members and other interested persons. Because long before the workshop Günther and Karin Sachs of CPM had outlined a time schedule comprising 35 precisely defined steps, this was possible.

The willingness of the authors to submit their manuscripts as hard copy and on diskette according to schedule — some already in July and August and some at the beginning of October 1991 — is greatly appreciated. This meant that the manuscripts of most of the main papers were available to all workshop participants as preparatory reading three weeks beforehand and that copy editing at Springer-Verlag had begun in mid-August.

By two weeks after the meeting Dr. Bondy had already processed, edited, and stored on diskette the tape recordings of the workshop discussion sessions. And so by the end of September a printed version of these discussions was passed on to the authors and editors. At the same time, copy editing of the discussions could begin at the publisher's.

For the copy editing thanks are extended to Sherryl Hirsch of Springer-Verlag, whose work and commitment in all regards was exemplary in accommodating the given schedule.

Finally, and above all, this book was able to be published so quickly thanks to the fact that the CPM staff used their desk top publishing system to process the manuscripts and discussions from diskette to reproducible page proofs.

And so — in addition to the scientific gain — the editorial work for the Second President's Workshop of the C.I.N.P. was a particularly pleasurable experience!

Most of all the C.I.N.P. and the editors are grateful to Günther and Karin Sachs not only for having the idea to produce the congress report so quickly, but also for putting it into effect, and who with their know-how had already organized the C.I.N.P. Congress in Munich in August 1988.

Once again, our special thanks to the CPM.

Munich, January 1992 Hanns Hippius

MIX
Papier aus verantwortungsvollen Quellen
Paper from responsible sources
FSC® C105338

If you have any concerns about our products,
you can contact us on
ProductSafety@springernature.com

In case Publisher is established outside the EU,
the EU authorized representative is:
Springer Nature Customer Service Center GmbH
Europaplatz 3, 69115 Heidelberg, Germany

Printed by Libri Plureos GmbH
in Hamburg, Germany